What Makes Humans Truly Exceptional

Larry Bell

What Makes Humans Truly Exceptional

Other books by Larry Bell:

Scared Witless: Prophets and Profits of Climate Doom
Climate of Corruption: Politics and Power behind the Global Warming Hoax
Cosmic Musings: Contemplating Life beyond Self
Reflections on Oceans and Puddles: One Hundred Reasons to be Enthusiastic, Grateful and Hopeful
Thinking Whole: Rejecting Half-Witted Left & Right Brain Limitations
Reinventing Ourselves: How Technology is Rapidly and Radically Transforming Humanity
Cyberwarfare: Targeting America, Our Infrastructure and Our Future
How Everything Happened: Including Us

STAIRWAY PRESS—Apache Junction

Book Cover Art: Albert Rajkumar
Cover Design by Chris Benson
www.BensonCreative.com

STAIRWAY≡PRESS
www.StairwayPress.com
1000 West Apache Trail, Suite 126
Apache Junction, AZ 85120

Dedication

Dedicated to my truly exceptional Sapiens family who endured, survived and accomplished much to make both the author and this book possible.

About the Author

LARRY BELL IS an endowed professor of space architecture at the University of Houston where he founded the Sasakawa International Center for Space Architecture (SICSA) and the graduate program in space architecture.

Larry's other recent books include: *How Everything Happened: Including Us* (2020), *Thinking Whole: Rejecting Half-Witted Left & Right Brain Limitations* (2018), *Reflections on Oceans and Puddles: One Hundred Reasons to be Enthusiastic, Grateful and Hopeful* (2017), *Cosmic Musings: Contemplating Life Beyond Self* (2016), *Scared Witless: Prophets and Profits of Climate Doom* (2015), and *Climate of Corruption: Politics and Power Behind the Global Warming Hoax* (2011). He is currently working on a new book co-authored with Buzz Aldrin, *Beyond Footprints and Flagpoles*.

Larry is also an entrepreneur who has co-founded several U.S. and international commercial space companies. One—established with NASA's Chief Engineer and two other partners—grew through mergers and acquisitions to employ more than 8,000 professionals, went on the New York Stock Exchange, and was purchased by General Dynamics.

Professor Bell's many national and international honors include two of the highest awards granted by Russia's Institute of Aeronautics and Cosmonautics: The Konstantin Tsiolkovsky Medal and the Yuri Gagarin Diploma.

Larry's name was placed in large letters on the Russian rocket that launched the first crew to the International Space Station.

PREFACE

THIS IS THE very first sentence and page of a yet unwritten personal book adventure. It will be a story about us—you and me—along with countless other truly exceptional Homo sapiens who made our lives possible: So it's about people who invented languages, cultures and cooperative societies; who sought spiritual and scientific understanding of the world and Universe; who created marvelous arts, architectures, tools and machines inconceivable to previous generations; and who continue to surprise us with future transformative discoveries and inventions that remain unimaginable today.

Having previously written several hundred articles on widely diverse topics for Forbes, Newsmax and other publications—along with numerous books—I seldom plan in advance what project will follow. With regard most particularly to books, I have learned to trust each writing and reflection experience to reveal special insights and puzzles that motivate me to reexamine, expand and explore previous and new vistas of discovery.

This most recent project began by deciding upon a title which asks—rather than answers—what I regard to be an enormously personal and complex series of questions.

An immediate and daunting challenge arises in defining what exemplary characteristics, values and achievements you and I might agree upon as being "exceptional." That term, above all, connotes very different and even mutually exclusionary meanings.

Applied as an adjective, the forming of an exception can be unusually good, bad or entirely neutral. Performance of a task can be far better than average, crimes can be unusually callous and cruel, or the weather can be uncommonly warm for January.

After all, lots of other contextually dependent words essentially convey similar meaning. For example: unsurpassed, outstanding, unprecedented, rare, out of the ordinary, surprising, unusual, atypical, anomalous, inconsistent, abnormal, strange, divergent, bizarre and deviant.

Weeding out adjectives and applications with clearly undesirable connotations still leaves us with little solid foundation to build on. Even applied as a compliment, referring to a concert performance as exceptional might mean that it was better than expected, or simply unusual.

So, let's first frame this exploration by contemplating which special qualities you and I might agree upon that exemplify "exceptional" characteristics, values and achievements that are truly worthy of guiding our highest aspirations.

Here again, the term "exceptionalism" connotes very different popular associations. Under special applications, it may refer to heroic, selfless deeds; famed abilities and accomplishments; and public recognition for uniquely influential acts or decisions.

Exceptionalism is also a quality that often goes quietly unnoticed and underappreciated by adoring throngs: something that each of us can aspire to and exhibit in the process of simultaneously enriching both our own lives and those of others. Viewed in this non-exclusionary context, exceptionalism is an inherently accessible human quality.

Although embraced as a fresh new start, this project consciously and intentionally revisits and reconnects with thematic pieces and patterns in an evolving mosaic of emerging perceptions and lessons explored most particularly in seven of my earlier books, along with insights from another co-edited with Apollo II astronaut Buzz Aldrin, which is in currently progress.

My 2018 book, *Thinking Whole: Rejecting Half-Witted Left & Right Brain Limitations,* addresses observations about important shared characteristics of very high-achieving individuals throughout history, including many I have been greatly blessed to know.

Whole brain thinking is evidenced in everyday life through awareness of surrounding environments. It is expressed through curiosity which compels our interests in how and why natural and man-made things work the way they do...interconnected relationships between ourselves and others...patterns and rhythms observed in nature...spiritual lessons and explorations that motivate higher purposes and values...inspirations experienced through forms, literature and stories of the past...music...everything combined that our whole minds can dare to contemplate, including ourselves.

As British-born neurologist and science writer Oliver Sacks instructed in a 2012 *New Yorker* magazine article:

> *To live on a day-to-day basis is insufficient for human beings; we need to transcend, transport, escape; we need meaning, understanding, and explanation; we need to see overall patterns in our lives. We need hope, the sense of a future. And we need freedom (or, at least, the illusion of freedom) to get beyond ourselves, whether with telescopes and microscopes and our ever-burgeoning technology, or in states of mind that allow us to travel to other worlds, to rise above our immediate surroundings.*

Sacks further observed that whole thinking often requires peaceful minds:

> *We may seek, too, a relaxing of inhibitions that makes it easier to bond with each other, or transports that make our consciousness of time and mortality easier to bear. We seek a holiday from our inner and outer restrictions, a more intense sense of the here and now, the beauty and value of the world we live in.*

And as American neurologist physician, philosopher and poet Debasish advises:

Let your thoughts, intentions, imaginations, and dreams fly under a clear blue sky with a spring breeze floating like a butterfly from flower to flower. See the beauty of mankind. Enjoy the nectar of life. It will shift your awareness to a brighter consciousness.

The ultimate benefit of thinking and living whole is realized through expanded and strengthened connections to life experiences. Writing in *The New Yorker*, David Brooks summed this up eloquently:

> *I've come to think flourishing consists of putting yourself in situations in which you lose self-consciousness and become fused with other people, experiences, or tasks. It happens sometimes when you are lost in a hard challenge, or when an artist or a craftsman becomes one with the brush or the tool. It happens sometimes while you're playing sports, or listening to music or lost in a story, or to some people when they feel enveloped in God's love. And it happens most when we connect with other people.*

Brooks concludes:

> *I've come to think that happiness isn't really produced by conscious accomplishments. Happiness is a measure of how thickly the unconscious parts of our minds are intertwined with other people and with activities. Happiness is determined by how much information and affection flows through us covertly every day and year.*

My 2016 book, *Cosmic Musings: Contemplating Life Beyond Self,* memorializes questions and observations that had occupied my mind-works over most of a lifetime, dating back to my earliest remembered thoughts. It reflects a personal quest to understand how my life—all life—fits into some larger order.

Cosmic Musings explores issues of spirituality, including concepts of morality and mortality, from broadly ranging religious and secular perspectives. Many of these ideas and ideals share common core values.

Most ancient Jewish scriptures, for example, are steeped in democratic precepts and moral concern which have had powerful influences upon other religions, including Christianity and Islam.

Mohammad and Buddha preached that there is no separation between the physical and spiritual world. The *Quran* instructs us to observe God's creation, work to understand those natural systems and constantly endeavor to seek knowledge.

Secular lessons have also advanced our understanding of natural and moral principles. Aristotle, Plato and Socrates are credited with establishing the foundations of modern philosophy and ethical reasoning in ancient Greece which later influenced the thinking of Saint Thomas Aquinas in the 13th century and Charles Darwin in the 21st.

Friedrich Nietzsche believed that values are relative and subjective, and that the exemplary human being must craft his/her own identity through self-realization without relying on transcending that life—such as either a God or a soul. From his point of view, every person should live his or her life for the sake of living it, not because of an anticipated afterlife (a concept he rejected).

Nietzsche viewed opposing psychological capacities that exist in everyone as great forces which must be balanced in order to live a full life. In his way of thinking, morality is a matter of reason and conscience, not tied to any religious faith.

Immanuel Kant saw things quite differently, believing that there can be no morality without religion. He theorized that moral thought has objective categories, differentiating between moral reasoning about how a person should act, which he referred to as "practical reason," vs. "pure reason" about something that exists.

Kant's categorical imperatives were perceived as universal in nature because they were there before being experienced. Accordingly, a person is duty-bound to act in a certain manner. Here, the action was more important than the outcome.

Kant supported what is now referred to as "foundations of

retribution" (or "just deserts"), meaning that a person who commits a crime should receive nothing more or less than a penalty called for. This idea prominently appears in most cultures and religions. Christians refer to it as the "Golden Rule," namely to "Do onto others as you would have them do onto us." Another translation to remember: "What goes around, comes around."

Richard Kinner and Jerry Kernes in the Division of Psychology in Education at Arizona State University-Tempe together with Phoenix counselor Therese Dautheribes (2000) compiled a "short list" of what they characterized as universal values drawn from well-known texts and documents of major world religions. Included are Judaism (the Tanaka), Christianity (the New Testament), Islam (the Quran), Hinduism (the Upanishads and the Bhagavad Gita), Confucianism (the Analects of Confucius), Taoism (the Tao Te Ching of Lao Tzu) and Buddhism (the Dhammapada).

The study also consulted with and reviewed materials of several secular organizations, including the American Atheists Inc., the American Humanist Association, and the United Nations Declaration of Human Rights.

Here's what they came up with:

- Commitment to something greater than self: to recognize the existence of a Supreme Being, higher truth, principle, transcendent purpose or meaning of existence; and to seek the Truth (or truths) and justice.
- Self-respect, but with humility and self-discipline: to care for oneself; to act in accordance with conscience, and to accept responsibility for one's behavior.
- Respect and caring for others: to recognize the connectedness between all people; to serve humankind and be helpful to individuals; to be caring, respectful, compassionate, tolerant, and forgiving of others; and to not hurt others (e.g., not murder, abuse, steal from, cheat or lie).
- Respect nature: to care for other living things and to protect the environment.

Purposeful living prompts each of us to constantly ask ourselves whether or not we have attempted to act in good faith to respect that list of universal values.

Achieving true exceptionalism doesn't really have anything to do with acquiring wealth or fame. Rather, it requires us to contemplate other more fundamental considerations. Am I consistently trustworthy to be a good spouse, parent and friend? Do I possess a generous nature where it comes to sharing credit and effort which recognizes the worth and contributions of others? Have I touched the lives of others in a constructive way simply because I could? Am I comfortable with that person that inhabits my own skin?

Here I'll again add another vital ingredient that fuels and personifies exceptionalism...the special "passion" word that sets humans apart from other perfectly wonderful creatures. Some might just as readily refer to this as "love."

Passion ratchets living whole up to a whole other level. It is what drives us to express our highest human potentials: to have empathy and truly care about others; to set goals; to meet challenges and seek excellence that sets exceptional examples; to create music, art and literature that lifts our intellect and spirits; and to believe in the power of worthwhile ideas and our God-given abilities to make them real. A well-lived life requires no less.

I first became motivated to record some of my own thoughts about some of these philosophical topics in a set of publicly unreleased essays that I wrote for my two sons when they were very young. The manuscript was later published in my 2017 book *Reflections on Oceans and Puddles: One Hundred Reasons To Be Enthusiastic, Grateful And Hopeful.*

While, as the title suggests, these reflections address topics which may generally be regarded as ranging from large to small, this is not intended to imply that bigger subjects are most important in any fundamental aspect. Marvels revealed by inner workings of a cell or atom are not necessarily of any less significance than those associated with the outer workings of the Universe.

Looking at our Earth home in a remote suburb of the Milky Way amongst millions of other galaxies we can imagine ourselves either insignificantly tiny or part of something unimaginably marvelous and immense.

Reflections mirrors my earliest remembered self-revelation both as a tiny elemental part and unlimited natural extension of those natural wonders.

I was a child, probably about seven or eight years old at the time, sitting quietly alone outside my house. The long Wisconsin winter had retreated, and grass had begun to appear amidst remaining patches of melting snow. The Sun was warm on my back, and the musky fragrance of wet earth was strong.

It suddenly dawned on me that I was directly connected to that magical world of my backyard. The materials that comprised the earth, grass, trees and even stones were the same stuff that I was made of. I wasn't just a spectator experiencing Nature. I was actually part of Nature, and somehow always would be, long after the trees were gone.

That awareness of being part of something really eternal and vast was very profound and exciting to me then, and it continues to be now.

We are all truly creatures of the natural Universe. The elements constituting every part of us and our beautiful blue planet have passed through other fiery stars, now gone, long before our mother Sun was born. Each of those elements, in turn, are comprised of atoms which are like tiny solar systems containing tremendous energy, each with a proton and neutron nucleus orbited by electrons that together direct the chemistry of life.

That Universe that spawned us is unimaginably vast and marvelous. Consider, if you will, that there are estimated to be more than 100 billion stars in our Milky Way galaxy, most which are larger than our "yellow dwarf" Sun. This spiral wheel is about 100,000 light years in diameter.

If we were somehow able to travel to a new star in our galaxy every hour of the day and night, it would take about six million years—much longer than humans have existed—to visit only about half of them.

Now also consider that there are estimated to be more than 100 billion galaxies in our known Universe. The Universe has existed about 13.8 billion years, about 9 billion years longer than our home planet, and is constantly changing, like a garden where new plants bud as others wither on a cosmic time schedule.

From our vantage point on a spiral arm of our galaxy, it is difficult to grasp the reality that those distant stars, and the planets and ghostly clouds that surround them, are part of our personal world. Humankind has a history of resisting any observations that placed us outside the center of the Universe. And in reality, the Universe probably has no center, except maybe a theoretical point where a Big Bang first set everything in motion. Yet there is really nothing to be upset about. The real estate we occupy has a wonderful location suited perfectly to our lives and a spectacular view.

Another difficulty many people seem to have with a cosmic perspective is that they feel it diminishes their significance. Perhaps you have seen graphic illustrations depicting the Milky Way galaxy, possibly printed on a tee shirt, with an arrow pointing to our solar system neighborhood with a note stating, "you are here." Some will take this as evidence that we are indeed small-timers living in the celestial boondocks. Others, however, may interpret this very differently, recognizing that we are all an integral part of something unfathomably majestic and empowering.

Yes, while it is true that our community and bodies are small relative to the Universe, we should remember that everything is small relative to the Universe. After all, does size really have anything at all to do with importance? Are boulders more important than hamsters? Is Saturn more important than the Earth? It seems to me that either everything is important, or nothing is. And that is purely a matter of personal decision.

The idea of a changing Universe can also be disquieting for some. If even planets, stars and galaxies are constantly being born and changing, ultimately only to die, then where does our human destiny lie? Upon what permanent ground can we build our spiritual refuge?

One answer is that change is the essential nature of life and spirit. Everything that we have the good judgment to enjoy is dynamic, revealing new dimensions of possibility with each transformation and discovery. The exciting news, if we accept it, is that nature is eternal, or as close to that as we dare imagine. And as manifestations of that wonderful condition, we are too.

My book, *How Everything Happened, Including Us* (2020), reviews the incomprehensible development and existence of our human lives, in fact all life, within a mind-numbingly expansive scope and time scale of historical events. Some occurred over billions of years ago, others in cosmic blinks; some were unfathomably large, others imperceptivity small; some had catastrophic impacts upon certain regions and species, benefiting others more fortunate; and some were willfully tragic, while others are spiritually inspiring.

In all cases, these events are now history. The results are self-evident. We can't change them.

The best we can do is attempt to draw constructive lessons regarding what and how things happened in the past to influence tomorrow's history.

Had things turned out quite differently, I wouldn't be here to write this book in my time-space, nor would you be reading this sentence now in yours.

The story—our story—begins a very long time ago, 13.7 billion years ago, about 5.3 billion before our planet was born 4.5 billion years ago. It then took about another 4 billion years for Earth to become teeming with simple, single-celled organisms that eventually evolved into you and me.

Within only the last ten thousand years some of those Homo sapiens ancestors of ours invented agriculture, battled and domesticated larger animals for food and clothing, competitively warred against each other and Neanderthal hunter-gatherers, established settlements, cities and empires, built great pyramids and cathedrals, formulated complex cultures and laws, developed advanced scientific methods and philosophies and composed inspirational literature, music and sonnets.

Some inventive and adventuresome Sapiens contemplated the architecture and workings of a celestial Universe and applied that knowledge to guide voyages of discovery, trade, conquest and migration to extend domains and dominions.

Others—within little more than the last century—have harnessed the power of lightning and atoms, have mastered flight, have travelled many times faster than the speed of sound, have transmitted information from everywhere to everywhere else via orbital satellites, have walked on the Moon and have conceived artifical brains that can already outsmart their human creators.

And both for better and worse, we're now really only getting started.

Reinventing Ourselves: How Technology is Rapidly and Radically Transforming Humanity (2019), observes that the concept of "thinking machines," which was first hypothesized during a 1956 meeting of scientists, mathematicians and engineers at Dartmouth College, is no longer theoretical. Powered by Artificial Intelligence (AI), the Internet of Things and the emergence of quantum computing, today's society is now experiencing the earliest beginnings of an inevitable, irreversible and unfathomably impactful information revolution.

Enthusiastic proponents promise tantalizingly optimistic visions: daily new conveniences previously conceivable only in the fertile imaginations of fiction writers but decades, or even a few years ago; personal living efficiencies and household economies that save precious time and money; enhanced mobility through shared on-demand transportation services that banish most private automobiles to rusty scrap heaps of oblivion; and safety from predatory behaviors of others through ubiquitous, ever-watchful interconnected security devices.

Others anticipate each of these technological enticements serving more as one-way pathways leading to ant farm societies.

And after all, if humans can invent machines which are increasingly smarter than we are, where does this lead? Are we in a sense "playing God" in a way that will render human reasoning obsolete? Can we integrate technological "thinking parts" into our biological anatomy to repair and replace failed sensory and motor response systems...just as we presently do with other organ and limb prosthetic devices?

The "new science" of quantum mechanics that supports an incredibly "smart" generation of computing capacity goes even so far as to suggest that our entire perception of the physical Universe exists only as illusory inventions of our individual minds. Whereas this concept presents a radical departure from traditional Western thought, it doesn't seem nearly so alien to much older Eastern philosophies.

Reinventing Ourselves, along with my other books noted,

reminds us that what makes people truly exceptional is our transformative ability to master our own lives and future.

Unlike our machines, each of us has inherited a vast natural fortune and capacity to know love and share compassion; to imagine and create marvelous unseen possibilities; to experience responsibilities and satisfactions of friendship and community; and against spectacular odds, to celebrate having won the bountiful lottery of life.

My book, *The Weaponization of AI and the Internet: How Global Networks of Infotech Overlords are Expanding Their Control Over Our Lives* (2019), emphasizes that while the human and social benefits of information technology exercised through free choices of the many are endless, controlled by special interests and interest agendas of an increasingly powerful few, the applications for social control and exploitation are equally boundless.

Remote surveillance and monitoring technologies follow and record us virtually everywhere—even in our own homes and cars. Ever-present smart phone "voice assistants" listen in on our private conversations and snitch on us to uninvited outsiders.

Electronic eavesdroppers catalog our special interests, analyze our psychological profiles and micro-target us for messaging. Personal viewpoints expressed in social media exchanges are monitored and censored by politically biased algorithms.

Whereas *Reinventing Ourselves* addressed numerous ways that AI-driven technologies are rapidly and radically impacting our lives, careers and social values, and *Weaponization* discusses consequential dangers of trading away personal freedoms for increased convenience and security, my book, *Cyberwarfare: Targeting America, Our Infrastructure, and Our Future* (2020), identifies and elaborates on terrifying national and personal security threats posed by weaponized warfare applications of information technologies.

Unlimited by geographic boundaries, state-sponsored cyberattacks move through cyberspace at the speed of light to simultaneously hit numerous diverse targets with devastating impacts. Purposes include infiltration and theft of information, disruption of critical energy grids, theft and extortion of money from financial networks and mass public disinformation campaigns.

Cyberwarfare has changed and escalated dangers and

stratagems of warfare in the post-nuclear age in fundamental ways. Until the cyber age came along, America's two oceans symbolized a reassuring sense of invulnerability. In addition, the threat of nuclear provocation was mediated by an uneasy balance of retaliatory capabilities between the two superpowers, the United States and Soviet Union, through common recognition of unacceptable risks and consequences.

An adversarial stand-off termed "mutually assured destruction," (or MAD), seemed effective not only because each knew the other possessed world-destroying power, but also because each had confidence in the well-proven efficacy of its weapons systems. This is no longer true.

Warfare conditions and risks in the cyber age have disrupted the strategic power balance of the MAD era as more than 100 nations, including many with modest financial and military resources, now possess substantial cyberwarfare capabilities

For all its unfathomably positive promises, the new information revolution has concomitantly spawned equally incomprehensible tools of self-destruction. Future human historians—should any survive—will determine whether our truly exceptional creativity led us to outsmart ourselves.

Table of Contents

WE'RE ALL BORN WINNERS

IF YOU TEND to be somewhat complacent about life, you might consider that you were born against incredible odds. Let's take just a few moments to review some statistics.

Your mother came into this world with about one million ova, of which only about 400 could have been expected to fully mature into eggs. Of those eggs, most waited decades for an opportunity to experience fertile fulfillment, only to be swept away in red floods.

Your father's contenders for self-actualization had chances which were far worse. Assuming that your dad was typical, he produced about 200 million sperm cells every day until numbers began to taper off at about the age of 45. If you were conceived when he was about 30 years old, we might imagine that about 1.5 trillion of those unfortunate little tadpoles fell by the wayside before they could enter anyone's gene pool.

When your opportunity as a sperm carrying half of your genetic identity came along, you faced awesome competition. There you were among 3,000 million or so would-be half-brothers and sisters waiting in the win-or-die Ovarian

Marathon. Timing for all contestants was critical. What if the egg wasn't waiting? What if one of your parents was away on a trip somewhere? What if your mother had a headache?

Of course, the results are now history. The race wasn't cancelled after all. Released with a mighty surge, you swam with heroic determination and were first to reach the finish line which hardened around you as a barrier to all others. You then shared your genetic treasures to begin a whole new life.

As remarkable as that victory was, it reveals only one episode among countless events of unbelievably good fortune that supported your success.

What if your father and mother had never met? Recognizing that the United States population alone is about 320 million people, they both had many other mates to choose from. Not only did their lives have to intersect at the same place and time, all conditions in their personal circumstances had to be compatible for that relationship to occur.

Similar coincidences (or divinely orchestrated plans) had to also occur in the lives of your grandparents. Your birth, in fact, depended upon a unique and unbroken chain of events involving ancestors dating back over the history of our human species. Conservatively assuming that this amounts to about 100 billion people so far, that's a fantastically lucky lottery ticket.

Those improbable events leading up to your birth may have limited meaning to those whose lives are not directly connected to your own. Beyond your spouse, your children and their children, ad-infinitum, most everyone else's life would probably go on pretty much in the same general way if you hadn't been born. If you hadn't arrived, no one would know the difference.

On the other hand, being born is the opportunity of a lifetime! It is an opportunity to let people appreciate what they would be missing without you. It is an opportunity to make good use of that genetic potential that you have won to experience life as fully as possible, and in the process, become your exceptional self.

You might remind yourself that being only one person in a general population of billions does not make you or them less special. This is easiest to remember when we put ourselves in the company of people who recognize they are born winners too. In accepting their own importance, they are likely to respect your value as well.

Joys of Being Ourselves

So just who are you? What personal qualities do you identify with that make you uniquely your exceptional self?

American psychologist and humanist philosopher Carl Rogers described self-concept as:

> ...the organized consistent conceptual gestalt composed of perceptions of the characteristics of 'I' or 'me' and the perceptions of the relationships of the 'I' or 'me' to others and to various aspects of life, together with the values attached to these perceptions. It is a gestalt which is available to awareness though not necessarily in awareness. It is a fluid and changing gestalt, a process, but at any given moment it is a specific entity.

However, as Austrian-born philosopher Ludwig Wittgenstein observed, just as the eye which is the source of the visual field but not in that visual field cannot see itself, so it is with the "I" which is the source of our consciousness.

In his 1984 book *Reasons and Persons,* British Oxford

Philosopher Derek Parfit observed:

> *When I believed that my existence was such a further fact, I seemed imprisoned in myself. My life seemed like a glass tunnel, through which I was moving faster every year, and at the end of which there was darkness. When I changed my view, the walls of my glass tunnel disappeared. I now live in the open air. There is still a difference between my life and the lives of other people. But the difference is less. Other people are closer. I am less concerned about the rest of my own life, and more concerned about the lives of others.*

Free your imagination now and try to recapture your earliest discoveries of what it must have been like inside your child mind before you realized that anyone else really existed.

That self-awareness became active long before you ever left the womb. As a fetus, you began to experience yourself through the consciousness of your mother's sudden movements and heartbeats that are amplified by the protective fluid that surrounded you.

Following birth, you began to be flooded with new experiences. That comforting heartbeat was replaced by louder, more varied sounds, including your own voice. Blurry objects appeared in your new field of vision, including your own hands and feet.

You also encountered other sensations for the first time, including tastes, smells and temperature changes. Some were uncomfortable, such as hunger and stomach gas. And you learned how to communicate your displeasure with your lungs, facial muscles, and tears to summon help.

It took you a while to differentiate yourself from other objects and events around you to know where you end and everything else begins. But you soon discovered that those things which are uniquely part of you could be directly controlled or experienced through your senses.

For example, you could move and make sounds on your own command. You could open your eyes to see...or close them to shut out light. You could feel things that touch you, but not things that touch someone else.

You began to recognize yourself as "you," distinctly separate from "them." As the center of your own world, you felt very special; so special in fact that everyone and everything outside yourself seemed to have been put there for your special benefit.

Later, you learned that this assumption is not entirely accurate. As you grew older, it became apparent to you that others also took their own existence and priorities seriously, and that to avoid problems, it was often a good idea to accommodate them.

Through interactions with the outside world you became more self-critical and introspective, and the in the process of growing, you accumulated information, acquired talents, developed relationships and recorded memories.

You compared yourself against prevailing models and expectations and tried your best to conform—even excel. Building upon these experiences, you continually molded and reshaped your self-concept and visible identity...or "personality," if you prefer to call it that.

Life is a voyage of self-discovery to meet a new person we are always in the process of becoming at each successive port along the way. Each leg of the journey offers challenges and lessons that change us.

Near-term concerns often press us to concentrate on the

present, and let future selves deal with tomorrow. For now, it may be all we can handle just getting to the next safe harbor.

So why dream about that exceptional person we want to eventually become, knowing that such illusions will probably be dispersed by unpredictable winds and tides of fate anyway?

One reason is because if we don't think about what we really want out of life, we won't have any basis for setting our course in the right direction. Another is because when we have some understanding of what we want, we can avoid wasting time and energy on routes destined to nowhere. A third is because understanding what we want can energize us to do something about it.

Dreaming about the future probably came easier in childhood. That was before setbacks and other disappointments challenged our confidence, before we forgot how special we are and before magic was disproven. Maybe our expectations began to wither when our parents and teachers warned us about how difficult it is to convert hopes into realities and stressed the importance of looking at life from a "practical point of view."

Predictably, years of experience have caused us to be much more discriminating in choosing goals and plans that temper adolescent idealism with mature rationality. In any event, we have learned to distinguish between idle fantasies and feasible possibilities, and between casual desires and worthwhile commitments...at least sometimes.

I won't argue that this hard-earned objectivity isn't necessary and beneficial in large part. But what if we go too far, allowing pragmatic conservatism to rule us completely?

Along with the innocence of youth, aren't we in danger of losing something else? Aren't we at risk of giving up those exciting visions of unbounded potentials that our child-selves naturally and wisely recognized? And if so, aren't we more inclined to abandon our expeditions of growth and discovery at readily accessible, yet marginal destinations?

Now, if you are truly brave, compare those observations with your life now. Are you happy and agreeable to be around most of the time? Are you missing out on many of the things you care most about? If so, is this OK? Is your life voyage moving in a direction that really interests you? If not, who is steering your ship?

Dreaming about the future may be a forgotten art we must relearn from our inner child-self. Of course, we also need to apply subsequent lessons of experience that help us understand what we value most in ourselves and our lives. Assuming that visions direct realities, and I believe they do, then dreaming is something that we can't afford to abandon or postpone. If we do, our ship of opportunity piloted by others may leave port without us.

Our Place in a Cosmic Continuum

From the time of our ancient ancestors, humankind has looked to the heavens for guidance and inspiration. The Sun, source of all life, governed seasons for hunting and signaled time to prepare shelter, clothing and food for winter's sovereignty. The Moon was a calendar for planting and harvesting as settlers asserted dominion over wilderness. The stars were maps and compasses that led explorers on voyages which extended territorial domains.

In these ways, and many more, the sky expressed the majesty and will of great powers that controlled human enterprises and destiny. It presented mysteries that raised human consciousness and stimulated self-reflection. It was the true home of the human soul where mortal achievements would ultimately find reward.

Advanced telescopes and space travel have brought some aspects of the heavens into closer view, yet mysteries of the

Larry Bell

human soul remain as elusive as ever. We now realize that the Moon, planets, Sun and stars are far more distant than our forefathers and foremothers perceived with unaided eyes.

In his 1986 book *The View from Nowhere*, Thomas Nagel asks:

> *How can I, who am thinking about the entire, centerless Universe, be anything so specific as THIS, this measly, gratuitous creature, existing in a tiny morsel of spacetime, with a definite and by no means universal mental and physical organization? How can I be anything so small and concrete and specific?*

Subscription to contemporary scientific theories often requires that we suspend conventional perceptions of what seems rational: like trying to imagine that the entire Universe of suns, planets and gerbils began as a singularity smaller than an atom that Big Banged its way out of virtual nonexistence, or conceiving of a time before time existed and gravity gave it a closed shape.

Albert Einstein's general theory of relativity, first published in 1916, provided a unified description of gravity as a geometric property of space and time (or "spacetime") which curves according to the energy and momentum of whatever matter and radiation are present. At the moment of the Big Bang, the boundary of spacetime, the equations of relativity broke down and classical physics no longer applied. A tiny fraction of a second later the entire "observable" Universe was no bigger than an atom, and temperature, density and curvature of the Universe all went to infinity.

Nevertheless, as MIT Professor Alan Guth points out:

12

What Makes Humans Truly Exceptional?

The Big Bang theory says nothing about what banged, why it banged, or what happened before it banged.[i]

In 1970, physicists Stephen Hawking and Robert Penrose proceeded to prove with mathematical certainty that the Universe must have begun as a "singularity." Hawking foresaw that quantum theory previously used to describe subatomic phenomena could then be applied to the Universe as a whole...the concept of quantum cosmology.

Hawking began his 1988 book *A Brief History of Time* with an anecdote he attributes to Bertrand Russell. As reported, during a public lecture on cosmology Russell was interrupted by an old lady in the audience who told him that everything he said was rubbish...she purportedly said:

The world is actually flat, and it's supported by a giant elephant that is standing on the back of a giant turtle.

When Russell asked what was supporting the turtle, she replied:

But it's turtles all the way down.

And perhaps very fertile turtles at that. After all, the word "cosmology" derives from Greek "kosmos" (Universe) and "gonos" (produce)...the same as "gonad." Cosmos and cosmetic also have the same Greek root for "adornment" or "arrangement."

And what if, as some scientists theorize, instead of a one-world perception whereby tiny blobs of particle presence appear to race around linearly in physical space with a

13

particular position and momentum, there are simultaneously two-worlds which allow those blobs to simultaneously take multiple paths.

If this seems outrageously "mysterious," does the mystery lie in why there are two different and contradictory sets of laws, or more fundamentally, why our traditional world view sense of reality doesn't comport with what is being observed? This is the case with quantum theory, which is yielding incredibly expanded quantum computing capacities that appear to defy basic principles of traditional physics and conventional "single universe" logic.

According to quantum theories, separate "parallel universes" can have a variety of optional spatial dimensions, operate simultaneously in different regions of spacetime and even manifest different alter egos of ourselves.

But if so, imagining that there are alter egos of each of us "out there somewhere," are they doing the same things now that we are? Is mine now writing this book, which you (from my perspective) are now reading later? And if not—assuming that our alter egos in parallel universes make different space-lifetime choices leading to alternate self-aware experiences— won't each alter ego set (or multiple sets) perceive their particular world realties quite differently than we do?

Philosophers and clerics have contemplated such mind-numbing mysteries long before contemporary theoreticians deduced Big Bang beginnings, quantum mechanics marvels and seemingly "preposterous" parallel universe propositions.

Plato's allegory, Phaedo, likens our very limited understanding of the objects and phenomena by which we perceive the world to emerging from a dark cave into sunlight where only vague shadows of what lies beyond that prison are cast dimly upon the wall. Those shadow forms of perception are both non-physical and non-mental, existing nowhere in time, space, mind or matter.

Aristotle argued that in order for animals to perceive, and for humans to reason, perfect copies rather than shadows of forms are required. He reasoned that the human mind can literally assume any form being contemplated or experienced and is also unique in its ability to become a blank state with no essential form.

Plato and Aristotle both influenced writings of Saint Thomas Aquinas during the early Middle Ages which have been integrated into Roman Catholic doctrine. Like Aristotle, Aquinas perceived the human being as a unified composite substance embodying two substantial principles: form and matter.

During the mid-17th century, the dualism philosophy of Rene Descartes envisioned the mind as a nonphysical substance capable of consciousness and self-awareness separate from the brain as the seat of intelligence.

In his *Meditations on First Philosophy*, Descartes determined that while he could doubt whether he had a body (it could be a dream or illusion created by an evil demon), he could not doubt that he had a mind. The mind was a "thinking thing" with the essence of himself which doubts, believes, hopes and contemplates. This led to his best-known philosophical statement "Cogito ergo sum", or "I think, therefore I am."

This suggests that while the human mind has a temporal and non-temporal part, T, the eternal non-temporal something continues to be able to retrieve a temporal something following the brain's death. Furthermore, the non-temporal something of the human mind which is somehow able to be retained or retrieved by the physical temporal substance can grow and expand during a human's lifetime.

Is there some aspect of "me-ness" and "you-ness" that gets recycled when that biodegradable part of our persona no longer

15

serves its purpose? Does history repeat ourselves? What about being reborn into a new and fresh body where we restart our mortal cycle? This general reincarnation concept has existed in various forms and religions for at least 3,000 years.

Ancient Greek philosophers wrote extensively about the concept of reincarnation, in connection with the legendary Orpheus and Pythagoras. Socrates apparently embraced the concept, writing:

> If all things which partook of life were to die, and after they were dead remained in the form of death, and did not come to life again, all would at last die, and nothing be alive—what other result could there be? ... Must not all things at last be swallowed up in death? ... But I am confident that there truly is such a thing as living again, and that the living spring from the dead, and that the souls of the dead are in existence, and that good souls have a better portion than the evil.[ii]

Plato taught that one's soul is immortal, preexists before birth, and is reborn many times. He argued that each soul chooses its next life, guided by its experiences in the previous lives. His student Aristotle initially accepted this idea, but later largely rejected the concepts of reincarnation and immortality.

Although belief in reincarnation is not held as a mainstream concept in Judaism and Christianity, it has been recognized within some of their followers. The Kabbalah body of teaching based upon an interpretation of Hebrew Scriptures includes this concept, and Hasidic Jews include it in their belief system as well.

Some early Christians, particularly the Gnostic Christians,

believed in reincarnation, as did some groups in southern Europe, at least until the Council of Constantinople in 553 A.D. Even today some Christians find support for reincarnation in the passage in the New Testament Book of Matthew in which Jesus seems to say that John the Baptist is the prophet Elijah returned.

In Hinduism, it is believed that an enduring soul spends a variable amount of time in another realm following death before becoming associated with a new human or animal body of either sex. Karma determines the conditions into which one is reborn based upon conduct during previous lives. Since life on Earth is considered less than fully desirable, an individual can engage in religious practices in each life until eventually earning release from the rebirth cycle. This results upon achieving union with the infinite spirit, nirvana, whereupon personal individuality loses meaning.

Similar to Hinduism, the natural law of karma, which is akin to laws of physics, determines circumstances of subsequent lives. Accordingly, there is continuity between personalities, but not persistence of identity. For this reason, Theravada Buddhists prefer the term "rebirth" to "reincarnation." Such circumstances of rebirths are not perceived as rewards or punishments handed out by a controlling God, but rather are natural results of various good deeds and misdeeds. The rebirth cycle involves innumerable lives including both sexes and nonhuman animals over many eons. This continues until all cravings are lost and nirvana is achieved.

In Theravada Buddhism found in southern parts of Asia, there is no enduring entity that persists from one life to the next. As emphasized in the doctrine of "anatta," or no soul, at the death of one personality a new one comes into being much like the flame of a dying candle can serve to light the flame of another. The new personality first emerges into a non-terrestrial plane of existence prior to becoming a terrestrial being.

Unlike Hindus and Buddhists, many religious populations in West Africa believe that individual rebirth is desirable and that life on Earth is preferable to that of the discarnate, limbo state. Some believe that individuals are generally reborn into the same family, and that souls may even split into several rebirths simultaneously. Included are "repeater children," in which one soul will harass a family by repeatedly dying as an infant or young child only to be reborn into the family again. Some groups also accept the possibility of rebirth into nonhuman animals, while others do not.

Many Inuit tribes, particularly those in northern and northwestern parts of North America, hold reincarnation concepts which vary greatly across different groups. As with certain West African groups, some believe that an individual can be reborn simultaneously as different people within the same family. Birthmarks corresponding to wounds of dead warriors are given as special credence. Depending upon the tribe, it isn't necessarily expected that all deceased individuals will be reborn, or that rebirth will necessarily assume same-sex or human form.

Various Shiite Muslim groups in western Asia, such as the Druses of Lebanon and Syria, and the Alevis in Turkey, believe in personal reincarnation without the Hindu or Buddhist concept of karma. Instead, God assigns souls to a series of lives in different circumstances that remain disconnected from one another until the ultimate Judgment Day when they are consigned to heaven or hell based upon the moral quality of their lifetime actions.

In Advayavada Buddhism, all reality is held to consist of only one substance, typically referred to as God or Nature, of which both material things and thought (e.g. body and mind) are matching attributes mirroring each other. Yet followers believe that what man and other sentient beings have in common with the rest of existence is not thought of any kind, but rather their "conatus." This conatus is an innate striving or drive to persevere successfully in all sentient life, of which sentient cognition is but one element. From a human perspective, conatus is experienced in the form of "progress," which is similar to Te, the "virtuous power" of the Tao in Taoism.

It somehow occurred to me in childhood that God, Nature, Universe, conatus or whatever we choose to name the mysteries and promises of all life, are metaphorically like a limitless and eternal ocean "soul." Like jellyfish, all creatures are formed from nutrients of that energetic life fluid.

As such, we are all natural spiritual extensions of a cosmic continuum and will continue to be so long as nature exists.

Beating the Odds against Life

It is impossible to even begin to comprehend the staggering odds against life existing on the particular planet we have the marvelous good fortune to inhabit, our own physically and intellectually complex lives very much in particular. For whatever reasons, our unfathomably rare celestial home came about in exactly the perfect neighborhood, with remarkable timing and exactly the right resources and environmental conditions.

But then again, if not for that myriad of incomprehensibly unlikely events which happened to occur at exactly the right places and times, we would not have any reason or capacity to wonder at all.

Complex life would not have evolved without both continuing change as well as stability. Had our Homo sapien ancestors arrived too early, suitable conditions for emergence and survival would not have been ready for them. If those essential conditions had not been sufficiently stable over the past 150,000 years, there would not have been enough time to reach that stage of being "us."

Nor can we ignore the significance of all-important location.

The potentially habitable real estate zone necessary to allow liquid water and metals for stable terrestrial planet formation within the spiral Milky Way galaxy is relatively narrow, extending from about two-thirds the distance from its center to periphery. The cosmic science jargon for metals includes all elements heavier than helium. Toward the galactic core, density and energetics become excessive for stable terrestrial planets. Moving outward from the habitable zone, too few metals exist for terrestrial planet formation.[iii]

Earth's habitable zone from the Sun is critical also to enable liquid water along with other vital conditions such as an oxygen atmosphere essential for complex life forms to exist. This region extends within a distance of 0.95 to 1.15 astronomical units (93 million miles per unit) around the Sun.

And although other atmospheres could support a lesser array of life forms, such as anaerobic bacteria, viable temperature range for Earth's aerobic life requires optimum levels of greenhouse gases containing carbon.[iv]

Plasma jets emitted from the Sun stormed the early Solar System with gale-force solar winds that swept away much of the light gases from inner planets. The outer planets were affected far less, which accounts for their high ice-gas nature.

Beyond critical size, planets hold an atmosphere. If too big, excessive pressure and temperature break inter-atomic and molecular bonds of matter. Structures in the Universe stabilize by a balance between opposing forces of nature. For example, a balance between gravitational and atomic forces exists when the matter's density is near that of the interior of a single atom.[v]

Among many fortunes allowing life on Earth, by astronomically good luck, both our Sun and planetary abode developed nearly optimum masses. The Sun is also a proper size to extend very long survival.

Whereas much smaller and dimmer stars occur far more commonly than our larger Sun, their meager outputs of energy would require life-supporting planets to orbit much closer than even Mercury does to our Sun. Whether such close life-sustaining orbits are feasible remains unknown. Unlike very high-mass stars which may survive as little as 300 years - not long enough for the likelihood of life to emerge - thrifty burning of hydrogen by small stars does offer the potential advantage of existing over trillions of years. Massive blue giants survive as little as 3 million years.[vi]

When small stars exhaust their hydrogen, they become red dwarfs rather than red giants. As they decrease in size, they become hotter, brighter and bluer. Earth's yellow dwarf Sun, on the other hand, will probably become a red giant which grows to engulf our planet within another 5 billion to 7 billion years.

Colossal early collisions with other space bodies brought about enormously beneficial consequences that gave rise to our living planet. One of these, a real doozy involving a collision with a former planet posthumously named Theia which occurred during the final stages of Earth's accretion, led to the Moon's creation.

Theia is estimated to have been perhaps half the size of Mars. Its impact is credited with accounting for Earth's 24-hour rotation and tilted 23-degree axial ecliptic obliquity which influences seasonal climate changes.

Earth's precession of the equinoxes in its orbit around the

Sun is influenced with respect to the Moon's distance. Exerting a stronger pull on the closest side of Earth, the Moon also causes monthly tidal variation. This cycle affects many aspects of life, exemplified by the calendar and menstrual cycles of many mammals.[vii]

Powered by energetics of Earth's molten core, our planet's dynamic plate tectonics architecture acts to moderate climate and to form and reformulate continents and coastal areas where land and sea life flourishes. Without plate tectonics, there would be no plant growth carbon cycle, and mountains would eventually erode away except for an occasional volcano.

Carbon dioxide and other greenhouse gases (gases containing more than two carbon atoms per molecule) must exist within a reasonable range to maintain viable climates.[viii]

CO_2 is also vital to plant growth. Green leaves use energy from sunlight through photosynthesis to chemically combine carbon dioxide from the air and water with nutrients trapped from the ground to produce sugars, which are the main source of food, fuel and fiber for life on Earth.

Together with early plant materials, nature has generously provided all the building blocks necessary to evolve complex organisms...us included.

Extraterrestrial sources of water and amino acids that occur in complex carbon molecules have arrived via a diverse spectra of high impact comet and meteorite deliveries. These, in turn, originated in giant molecular clouds in the Milky Way which contain organic materials as well. Included are free atoms of hydrogen and helium along with a wide variety of carbon-containing molecules such as carbon monoxide, formaldehyde and alcohol. Under appropriate conditions these organic materials react to form amino acids—essential components for proteins of life.[ix]

Although our human existence is very recent in

cosmological terms, it took a great deal longer than our minds can fathom for Earth's conditions to be amenable to our arrival. That story dates back about 3.7 billion years ago to the first simple, single-celled organisms that eventually evolved into you and me.

It took another 800 million years for those animated self-replicating microdots to transform into such complex living marvels as ferns, flowers, cephalopods, dinosaurs, reptiles and a branch of vertebrates we now call mammals. A sub-branch of some of those very lucky surviving creatures eventually grew brain frontal lobes capable of self-awareness and tribal social behaviors. About 300,000 years ago, a single branch of these "primates" underwent a genetic mutation that allowed speech.

Although our ancestral Hominid line has been around longer, about 2.5 million years, we're youngsters compared to the age of the dinosaurs which existed some 160 million years. On this cosmic time scale, modern humans developed very recently—roughly 160,000 years ago. Our existence relative to Earth's age is infinitesimal, amounting to only 0.000375 percent of our planet's existence, or about 0.0000107 percent of the Universe's existence.[x]

Having progressed this far, are we intelligent beings alone "out here?" Stephen Hawking commented on this, observing:

> I believe alien life is quite common in the Universe, although intelligent life is less so. Some say it has yet to appear on planet Earth.

So, let's first assume that by "intelligent" life we're referring to creatures we might be able to engage in meaningful communications if not actually enjoy socializing with; particularly if they don't pose a threat to self-esteem and wellbeing. Well according to a famous "Drake equation," the probable number of such communicating civilizations could be quite enormous.

Formulated by Frank Drake in 1961, the Drake equation speculates probabilities regarding the number of planetary civilizations by factoring in such considerations as: the rate of star formation; the fraction of those stars with habitable planets; the fraction of those planets which might develop life; the fraction of intelligent life; the further fraction of those capable of developing detectable technology; and finally, the length of time such civilizations might be detectable. The fundamental problem here is that sound statistical estimates are rendered impossible since our own circumstances offer the only known (and highly biased) example.

Those estimates vary widely. In 1966 Carl Sagan optimistically suggested that there might be as many as one million communicating civilizations in the Milky Way alone, although he later decided that number might be far smaller. Mathematical physicists Frank Tipler and John D. Barrow—far more pessimistically—put their average-per-galaxy guess at less than one.

In any case, more than 50 years of SETI radio telescope searches haven't turned up any signs of intelligent alien presence yet...no evidence that our galaxy is teeming with powerful transmitters other than our own continuously broadcasting and receiving near the 21cm hydrogen frequency as we hoped to find.

Frank Drake described his own equation as just a way of "organizing our ignorance" on the subject. Still, as a general roadmap of what we need to learn in order to understand this existential question, it has formed the backbone of astrobiology as a science.

According to a "Fermi-Hart paradox," it looks like the odds of finding intelligent life on other Earth-like planets aren't great due to the long time required for such beings to evolve coupled with a relatively short remaining life span of our planet when the Sun brightens as predicted in another billion years or so. Physicists Enrico Fermi and Michael Hart argued given that our Sun is a typical and relatively young star while billions of stars in the galaxy are billions of years older, we should have seen some evidence of advanced alien civilizations by now.

Hence Fermi's question: "where is everybody?"

Disasters and Do-Overs

Were it not for exceptionally good fortune, we wouldn't be here either.

The world's ecosystems experienced some big setbacks along the evolutionary highway. Five separate mass extinctions attributed to major climate changes, catastrophic asteroid impacts and massive volcanic eruptions variously killed between three-quarters and 96 percent of all life species.

Many of those species that perished would likely have evolved into creatures far different from us, perhaps competitively blocking our emergence and development of human intellect and innovation altogether. Consider, for example, how enormously lucky we are to have survived—even benefited from—the catastrophic meteorite strike of the scale that exterminated dinosaurs.

We can also be very grateful that our since-evolved lives are made much safer from similar events thanks to cosmic debris vacuumed up by Jupiter's powerful gravity.

By "life," molecular biology indicates that all plants and creatures that exist today originated in an oxygen-free environment around 3.5 billion years ago from a common single-celled organism called "LUCA" (last universal common ancestor).

It then took nearly another three billion years for enough

of that atmospheric oxygen to accumulate that enabled the first complex multi-cellular reproducing life forms to "recently" develop little more than 530 million years ago.

Referred to as the "Cambrian explosion," fossils preserved in the Burgess shale of British Columbia reveal the appearance during this period of even more primitive animal phyla with very different body plans than exist in today's biota. Remarkably, this phenomenon emerged over the short time space of a million years or less—a geological eye blink.[xi]

Reptiles first appeared in the late Triassic period about 300 million years ago during what is referred to as the Mesozoic Era, a time of high oxygen levels. Some 230 million years ago they developed a new respiratory system with small, rigid sac-like appendages called a septate added to old lungs to handle lowering oxygen levels.

This change led to the appearance of dinosaurs because this more efficient oxygen-handling system enabled them to survive the mass extinction at the end of the Triassic. The more advanced lungs also supported dinosaur evolution into birds. Lizards, on the other hand, didn't acquire this evolutionary advantage, which is why they cannot thrive in high altitude thin air.

Birds, the surviving descendants of dinosaurs, still have lungs of dinosaur structure. These air sacs are filled with air when birds breathe, where it is stored a short time before passing to their lungs for the next inhalation. The highly efficient exchange between air and blood enables birds to extract far more oxygen than animals of comparable size. This explains why geese can fly over the Himalayas at altitudes that are lethal to humans.[xii]

Many paleontologists believe that periodic ups and downs in oxygen levels have affected much of the previous and existing extreme variation life forms.

About 542 million years ago, atmospheric oxygen was lower than today and fluctuated for the next 100 million years. Afterward, oxygen rose steadily until 400 years ago. It was some 25 percent at the beginning of the Devonian period. Then a steep decline occurred, followed by a rise peaking near the end of the Carboniferous period when oxygen levels are believed to have been near 30 percent. This was again followed by a nadir at about 12 percent oxygen at the Triassic end—causing a mass extinction—followed by an irregular rise to today's 21 percent level.

Mass extinctions at intervals of millions of years since that time have had major evolutionary influences, eradicating some life forms and providing advantages for the reemergence of new species to take places of those lost. Modern humans are very much included among the big beneficiaries.

There are likely to be many individual events, including asteroid impacts, climate changes, reformation of continents and toxic gas releases from volcanic eruptions. Some factors remain highly speculative with many unanswered questions.

Scientists have discovered at least five mass extinctions when anywhere between 50 percent and 75 percent of life were lost. These "Big Five" are summarized from oldest to most recent in the following.[xiii]

The Ordovician-Silurian Extinction (439 million years ago):

An Ordovician event that began around 439 million years ago wiped out an estimated 86 percent of all species on Earth. Key theories attribute the extinctions to two major causes: glaciation and falling sea levels.

It is believed that Earth was covered at the time with a vast quantity of plants that removed so much carbon dioxide from the air that temperatures fell drastically. Falling sea levels resulting from the formation of the Appalachian mountain range may have taken a toll on the majority of early marine animals.

The Late Devonian Extinction (between about 349-364 million years ago):

Although not known whether this extinction occurred as a single brief event or one spread over hundreds of thousands of years, it may have terminated as many as 75 percent of all species. Some studies attribute the cause to giant land plants with deep roots that released nutrients into oceans that resulted in algal blooms which depleted the waters of oxygen, thus, animal life. Another contributing factor may have been volcanic ash that cooled the Earth's temperature, killing off spiders and scorpion-type creatures that had emerged on land by this time.

An ancient amphibian cousin, the elpistostegalians, that had also made it to land became extinct, and vertebrates didn't appear again on land until about 10 million years later. Land animals became common again only after the early Carboniferous period 345 million years ago. This second land colonization correlates with oxygen levels close to today's levels.

Had the late Devonian extinction not occurred, these ichthyostegalians from which we all evolved might not have happened, and we humans would not exist today.

The Permian-Triassic Extinction (about 251 million years ago):

Considered the worst in history, this extinction terminated an estimated 96 percent of all species including ancient corals.

Today's corals are an entirely different group.

Often referred to as "The Great Dying," it is believed to have been caused by an enormous Siberian volcanic eruption that filled the air with carbon dioxide which fed bacteria that emitted huge amounts of methane.

According to theory, the Earth warmed, oceans became acidic, and life descended from only the four percent of the surviving species. Following the event, complex marine life rapidly developed, and snails, urchins and crabs emerged as new species.

The Triassic-Jurassic Extinction (between about 119-214 million years ago):

This event may have killed off as many as 80 percent of species that had survived the previous Permian-Triassic extinction, opening a path forward for the evolution of dinosaurs that later existed for around 135 million years. A key cause is broadly attributed to an asteroid impact.

Although small mammals outnumbered dinosaurs during the beginning of this era, by the end, dinosaurs' ancestors (archosaurs) reigned supreme.

The Cretaceous-Paleogene Extinction (about 65 million years ago):

The most famous of the Big Five extinctions brought on the extinction of dinosaurs along with an estimated 76 percent of all other life on Earth including many mammals. Attributed to a combination of a ten-kilometer asteroid impact in the Yucatan and volcanic activity, it allowed for the evolution of new mammal species and fish such as sharks at sea.

An asteroid of similar size is expected to impact Earth on average every 50 million to 100 million years. Huge impact craters exist in Woodleigh, Australia and Manicouagan, Quebec, Canada. The latter impact may have occurred 200 to 250 million years ago, perhaps also aiding the Permian-Triassic extinction.

During the Cretaceous period, many shallow seas flooded parts of continents leading to extensive marine sediment formation. Because fossil preservation occurs in sedimentary rocks, a better record of Cretaceous life exists than for other mass extinctions.

THE BIRTH OF INTELLIGENCE

GEOLOGIC TIMESCALES SHOW enormous time required for events that led to Earth's present biotic distribution. The period before the Cambrian, (the Precambrian), lasted about three billion years—approximately 80 percent of Earth's existence. It took that long before there was sufficient atmospheric oxygen to enable the evolution of complex life.[xiv]

The once rather peaceful pre-Cambrian seas became transformed into a liquid jungle of newly-mobile hunters and hunted. And while some animals (such as sponges) lost their nerve cells and regressed to a vegetative life, others, especially predators, evolved increasingly sophisticated sense organs, memories and vegetative minds.

In his last book, *The Formation of Vegetable Mould, Through the Action of Worms,* Darwin wrote that worms, which can distinguish between light and dark and modulate their responses to threats suggests "the presence of a mind of some kind."

Darwin noted that worms generally stay underground safe

from predators during daylight hours, and while they have no ears, are very sensitive to vibrations conducted through the earth such as footsteps of approaching animals. All of these sensations, Darwin concluded, are transmitted to collections of nerve cells (he called them "the cerebral ganglia") in the worm's head.

Following an 1859 discovery by Louis Agassiz that the jellyfish Biugainvillea had a substantial nervous system containing about a thousand nerve cells, George John Romanes—Darwin's young friend and student—demonstrated in 1883 all that the jellyfish employed both autonomous, local network-dependent mechanisms and centrally-coordinated activities through the circular brain-like organ that ran along margins of the bell.

Romanes wrote:

> *[N]erve tissue Is invariably present in all species whose zoological position is not below that of the Hydroza. The lowest animals in which it has hitherto been directed are the Medusae, or jellyfishes, and from them upwards its occurrence is, as I have said, invariable.*
>
> *Wherever it does occur, its fundamental structure is very much the same, so that whether we meet with nerve tissue in jellyfish, an oyster, an insect, a bird, or a man, we have no difficulty in recognizing its structural units as everywhere more or less similar.*[xv]

Although perhaps not very smart, jellyfish do seem to have minds of their own. For example, they can change direction and depth, and many have a "fishing" behavior that involves turning upside down for a minute, spreading their tentacles like a net, and then righting themselves, which they do by virtue of eight gravity-sensing balancing organs. Box jellyfish (Cubomedusae) have fully developed image-forming eyes, not so different from our own.

At the same time that Romanes was vivisecting jellyfish and starfish, a passionate young Darwinist named Sigmond Freud was studying cells of vertebrates and invertebrates in the Vienna laboratory of psychologist Ernst Brucke. Freud was particularly interested in comparing a very primitive vertebrate (Petromyzon, a lamprey) with those of an invertebrate (a crayfish).

Freud produced meticulous, beautiful illustrations showing all nerve cells were found to be basically similar... including with those of human beings. The nerve cell body and its processes—dendrites and axons—constitute the common basic building blocks, serving as each nervous system's signaling units.

Although neurons may differ in shape and size, they are essentially the same from the most primitive animal to the most advanced. Only their number and organization significantly differ. Whereas we have a hundred billion nerve cells, while a jellyfish has a thousand, their status as cells capable of rapid and repetitive firing is essentially the same.[xvi]

The crucial role of synapses—the junctions between neurons—to modulate individual organism behaviors was clarified only at the close of the nineteenth century by the great Spanish anatomist Santiago Ramon y Cajal. English neurophysiologist Charles Sherrington coined the word "synapse," demonstrating that synapses could be excitatory or inhibitory in function.

Within a few years of Darwin's death, it had been discovered that even single-celled organisms like the protozoa could exhibit a range of adaptive responses. A tiny, stalked, trumpet-shaped unicellular "Stentor" organism employs a repertoire of at least five different responses to being touched before finally detaching itself to find a new site if these basic responses are ineffective.

If touched again, the Stentor will skip the intermediate steps and immediately take off for another site. In this sense, it has become sensitized to "remember" an unpleasant experience and learn from it, although the memory lasts for only a few minutes. [xvii]

In the 1960s, Eric Kandel embarked on a study of the cellular basis of memory and learning based upon examination of a giant sea snail, the Aplysia. The mollusk was selected because it has a relatively few (20,000 or so) neurons which are distributed in ten or so ganglia of about 2,000 neurons apiece. Aplysia has particularly large neurons—some that are visible to the naked eye—connected with one another in fixed anatomical circuits.

Kandel wrote:

> *I appreciated that all animals have some form of mental life that reflects the architecture of their nervous system.*[xviii]

The ancient Aplysia exhibits a protective reflective reflex that, when threatened, withdraws its exposed gill to safety. By recording and sometimes stimulating nerve cells and synapses in the abdominal ganglion that governs these responses, Kendel was able to show that it exhibited both short- and long-term memory. He observed that the short-term memory, as involved in habituation and sensitization, depended on functional changes in synapses. Longer-term memory, which might sometimes last several months, evidenced structural changes in the synapses. In neither case was there any change in the actual circuits.[xix]

Whereby Aplysia has only about 20,000 neurons distributed in ganglia throughout its body, an insect, despite tiny size, may have up to a million nerve cells which enable extraordinary cognitive feats. As previously discussed, bees are expert in recognizing different colors, smells and geometrical shapes presented in a laboratory setting, as well as systematic transformations of these. In addition, they not only recognize the appropriate patterns and smells and colors, but can also explore and remember their locations, and communicate these coordinates to fellow bees.

Laboratory observations have demonstrated that members of a highly social paper wasp species can even recognize individual faces of others. Such face-learning cognitive skill have been broadly associated only with mammals.

As for exhibiting any conclusive definition "true consciousness," the very term begs clear interpretation. For example, Charles Darwin noted in *The Voyage of the Beagle* that an octopus in a tidal pool seemed to interact with him in a watchfully curious and even playful manner. As also observed by others, domesticated cephalopods often seem to evoke a similar sense of empathy and emotional proximity with humans that we associate with consciousness of felines and dogs.[xx]

Nature has employed at least two very different ways of making a brain—indeed, there are almost as many ways as there are phyla in the animal kingdom. Yet as Oliver Sacks reminds us, mind, to varying degrees, has arisen or is embodied in all of these, despite the profound biological gulf that separates them from one another, and us from them.

Modern humans have inherited a large neurological advantage. Our brains containing upwards of a hundred trillion neurons, each with up to ten thousand synapses, afford practically infinite capacities for conscious and unconscious manipulations.

Unfathomably, each of these neuronal signaling groups remain in constant communication with each other, weaving, many times a second in continuously changing but always meaningful patterns.

My friend David Eagleman, an adjunct professor of neuroscience at Stanford University, offers an analogy for how this all comes together into higher consciousness by broadly comparing the process to swarm intelligence expressed by an ant colony. Although the colony as a collective whole accomplishes extraordinary feats, each ant individually behaves simplistically. It just follows simple rules.

The queen doesn't give commanding orders; she doesn't coordinate the behavior from on high. Instead, each ant reacts to local chemical signals from other ants, larvae, intruders, food, waste, or leaves. Each ant is a modest, autonomous unit whose reactions depend only on its local environment and the genetically encoded rules for its variety of ant.

When enough ants come together, a superorganism emerges with collective properties that are more sophisticated than its basic parts. This phenomenon, known as "emergence," is what happens when simple units interact in the right ways so that something larger arises.

And so, it goes with the brain. A neuron is simply a specialized cell, just like other cells in your body, but with some specializations that allow it to grow processes and propagate electrical signals. Like an ant, an individual brain cell just runs its local program its whole life, carrying electrical signals along its membrane, spitting out neurotransmitters when the time comes for it and being spat upon by the neurotransmitters of other cells.

Eagleman concludes:

That's it. It lives in darkness. Each neuron spends its life embedded in a network of other cells, simply responding to signals. It doesn't know if it's involved in moving your eyes to read Shakespeare, or moving your hands to play Beethoven. It doesn't know about you. Although your goals, intentions, and abilities are completely dependent on the existence of these little neurons, they live on a smaller scale, with no awareness of the thing they have come together to build.[xxi]

Our Consciousness of "Self"

So, with all those robotically programmed ant-like neurons why aren't we all just wondering around like mindless zombies? Even more, what makes each of a uniquely self-consciously-identifying "me"?

In asking the question *"who am I,"* David Eagleman reportedly answers himself, saying:

When I think about who I am, there's one aspect above all else that can't be ignored: I am a sentient being. I experience my existence. I feel like I'm here. Looking out on the world through these eyes, perceiving this Technicolor show from my own center stage. Let's call this feeling consciousness or awareness.[xxii]

Eighteenth century German philosopher Emanuel Kant referred to this self-awareness as an unknowable "transcendental apperception of the ego"—seeing this self not as something perceived by the senses, but rather as something spiritually eternal which transcends us. In his 1781 book *A Critique of Pure Reason,* he maintained that the mind relies upon "a priori forms," abilities and ideas supplied by divine acts which were always within us, and from which everything else flows.[xxiii]

Princeton neurosciences Professor Michael Graziano also contemplates relationships between our physical brains and transcendental thoughts. In his book *Consciousness and the Social Brain,* he broadly defines this personal consciousness of self as the window through which we understand:

> *The essence of self-awareness...the spark that makes us us...something lovely that apparently is buried inside us that makes us aware of ourselves and the world.*[xxiv]

Australian-born philosopher Ludwig Wittgenstein observed, just as the eye, which is the source of the visual field but not in the visual field, cannot see itself, so it is with the "I" which is the source of our consciousness.

But still, how did each of us come to experience that conscious "me-ish" awareness that sets us apart from everyone else? For example, why is it that we can feel things that touch us, but not things that touch someone else?

American psychologist and humanist philosopher Carl Rogers described such a self-concept as:

> *...the organized conceptual gestalt composed of the characteristics of 'I' or 'me' and the perceptions of the relationships of the 'I' or 'me' to others and to various aspects of life, together with the values attached to these perceptions. It is a gestalt which is available to awareness though not necessarily in awareness.*

What Makes Humans Truly Exceptional?

It is a fluid and changing gestalt, a process, but at any given moment is a specific entity.[xxv]

So yes, each of us somehow experience our own gestalt, which in turn is somehow part, and perhaps in some ways also independent of an unknowably incomprehensible large and timeless gestalt. Some may refer to this as God, others as nature and maybe still others as a quasi-mechanical or metaphysical phenomenon.

Nevertheless, we don't necessarily have to understand how things work—or why—in order to marvel at the fact that they exist. Descartes, in his *Meditations on First Philosophy,* appreciated that while he could doubt that he had a body (it could just be a dream or illusion created by an evil demon), he could not doubt that he had a mind...a "thinking thing" which carried the essence of himself which doubts, believes, hopes and contemplates.

V.S. Ramachandra, author of *The Tell-Tale Brain: A Neuroscientist's Quest for what makes Us Human,* marvels, as we all should, about what makes our high level of human consciousness possible. He asks:

How can a three-pound mass of jelly that you can hold in the palm imagine angels, contemplate the meaning of infinity, and even question its own place in the cosmos?

Especially awe inspiring, is the fact that any single brain, including yours, is made up of atoms that were forged in the hearts of countless far-flung stars billions of years ago. These particles drifted for eons and light years until gravity and change brought them together here, now.

These atoms form a conglomerate—your brain—that can not only ponder the very stars that gave it birth but can also think about its own ability to think and wonder about its own ability to wonder. With arrival of humans, it has been said, the Universe has suddenly become conscious of itself. This, truly, is the greatest mystery of all.

How We Sapiens Got Big Heads

Suzana Herculano-Houzel notes in her book *The Human Advantage* that as presumptuous as it may seem, it is a fact that we are the only species to study itself and others to generate knowledge that transcends what is observed firsthand; to tamper with itself, fixing imperfections with the likes of glasses, implants, and surgery and thus changing the odds of natural selection; and to modify its environment so extensively (for better or for worse), extending its habitat to improbable locations:

> *We are the only species to use tools to make other tools and technologies that extend the range of problems it can tackle; to further its abilities by seeking harder and harder problems to solve; and to invent ways to register knowledge and to instruct later generations that go beyond teaching by direct demonstration. Even though all this may be achieved through no particular cognitive ability exclusive to our species, we certainly take these abilities to a level of complexity and flexibility that is rivaled by none.*[xxvi]

What Makes Humans Truly Exceptional?

How did this marvelous cognitive engine that only emerged a few millennia ago develop to drive the evolution of technology, language and complex social behavior? Carl Sagan attributed its origin to a common ancestral proto-reptile where somewhere in the steaming jungles of the Carboniferous period there emerged an organism that for the first time in history of the world had more information in its brains than in its genes. It was an early reptile which, were we to come upon it in these sophisticated times, we would probably not describe as exceptionally intelligent.

> But its brain was a symbolic turning point in the history of life. The two subsequent bursts of brain evolution, accompanying the emergence of mammals and the advent of manlike primates, were still the important advances in the evolution of intelligence.[xxvii]

Whereas elementary life has existed on this planet for close to four billion years, the first hominids emerged in East Africa on a rapid evolutionary growth path from an earlier genus apes called *Australopithecines* (meaning "Southern Ape") only as recently as about 2.5 million years ago.

Whereas the idea that man could be regarded as a mere animal—an ape—descended from other animals has provoked outrage and ridicule by some, Charles Darwin's theory of evolution through "natural selection" based directly upon observed common and differentiated features of geographically isolated living species continues to be validated by fossil discoveries and DNA genetic sequencing revelations.[xxviii]

Discovery of the structure of DNA by James Watson and Frances Crick which gave evolution emphatic validity has also enabled detailed understanding of the human genome, genetic

41

diseases and disorders and most appropriate therapeutic treatments.[xxix]

Genes are specific sequences of chemicals called "nucleotides" which account for physical variations among people such as skin, eye and hair color. About three billion nucleotides reside within each of twenty-three chromosomes contained in every human sperm cell or egg. Whereas chromosomes among people are largely the same, only one in about each 1,000 nucleotides show a difference.

So far, only one gene is known to have survived unchanged in the biosphere since 1 billion years ago—histone$_4$ codes for DNA molecule housing. All others have and will continue to change. And yet, because of that great evolutionary engine of natural selection, every species became unique, just as each individual is also.

Evolution's implication is that all life ascended from a common single-celled ancestor through a bottom-up, self-organization, coevolution with other life forms, sometimes by trial and error and sometimes interceded by accidents. This "whatever works best" concept along with the projected billions of years required for higher forms to emerge was even more difficult to imagine in brains conditioned to short-term thinking during Darwin's time.[xxx]

By about 55 million years ago, during the Eocene period, there was a great proliferation of primates, both arboreal and ground-dwelling, and the evolution of a line of descent that eventually led to man. Based upon endocranial fossil evidence, one a prosimian called *Tetonius* exhibited in tiny nubs where frontal lobes later evolved.

The first fossil evidence of a brain of even vaguely human aspects dates back to eighteen million years to the Miocene period, when an anthropoid ape called *Proconsul* appeared. Proconsul was probably ancestral to the present great apes and possibly also to modern humans as well.

DNA evidence indicates gorillas severed from our lineage 7 to 9 million years ago, while chimpanzees and bipedal apes went their separate ways 5 to 7 million years ago. The gap between bipedal apes and humanoids is the notorious missing link.[xxxi]

In July 2002, a French paleoanthropologist named Michael Brunet and his team found a complete, seven-million-year-old Australopithecines cranium with canine teeth in Chad's Djurab Desert in Central Africa which they named Sahelanthropus tchadensis. The site was 1,500 miles west of the East Africa Rift Valley and South Africa, where previous searches for early hominids had been concentrated.

Of special interest, the partial skull showed a modification in the direction of humans with its lower face projecting less than apes. Brunet concluded that the newly discovered species might be a close relative to our common ancestor with chimpanzee—the earliest member of human lineage. Characteristic of discoveries, Brunet's contention did not meet the science community without dissent.[xxxii]

Although it is debatable by some whether Australopithecines were directly ancestral to modern humans or merely close collateral relatives, casts from their fossil skulls reveal that like us, they possessed large brains for their body weight. It is certain, however, that a great abundance of apelike animals existed by about five million years ago.

Some Australopithecines clearly not of the genus Homo (not human) were still incompletely bipedal with brain masses only about a third of the size of the average adult modern human. Were we to meet an Australopithecine today, we would perhaps be struck by an almost total absence of forehead.

A French-American team found an ape-like-faced Australopithecus fossil with small canine teeth nicknamed

"Lucy" (LUCA, Our Last Universal Common Ancestor) that lived in Ethiopia and Kenya roughly between three and 3.7 million years ago. Although bipedal, the species had long arms to enable adept tree climbing and chipped stones for rudimentary tools.

The Australopithecines robustus possessed impressive "nut-cracker" teeth and a remarkable evolutionary stability. Its endocranial volume varied very little from specimen to specimen over millions of years of time. This species was taller and heavier than the Gracile Australopithecines variant that walked on two feet and had brain volumes of about 500 cubic centimeters, some 100 cubic centimeters larger than brains of the modern chimpanzee.

Australopithecines gracile appear to be a considerably older species, with much more variance in endocranial volumes over time than their robust cousins. Judging again from their teeth, the Gracile probably ate meat as well as vegetables.

The A. Gracile fossil sites reveal implements made of stone and animal bones, horns and teeth which were painstakingly carved, broken, rubbed and polished to make chipping, flaking, pounding and cutting tools. No such tools have been associated with the A. Robustus species, whose brain to body weight was only about half as large as the A. Gracile.

It is natural to speculate that this large relative difference in brain size may have a connection between having tools or no tools. Paleontologists also theorize that bipedalism may have preceded "encephalization," by which our ancestors and cousins walked on two legs before they evolved big brains.[xxxiii]

A remarkable aspect of the archaeological record concerning tools is that as soon as they appear at all they appear in enormous abundance. This suggests the existence of ancient stone craft education programs which passed on the skills between tribes from generation to generation.

Less than three million years ago, the glacial-interglacial times of the Pleistocene Epoch emerged, whereby climate changes caused both chimpanzee and australopithecine populations to become downsized and forced into isolated areas in order to sustain traditional ways of life. This isolation may have enabled a new variant of australopithecine to emerge about 2.5 million years ago—the indisputable beginning of the Homo lineage having bigger brains with cortical folds.[xxxiv]

African forests shrank, influenced by a long, cool and dry period. And while some species adapted, others were forced to develop new food sources. By three million years ago, some Australopithecines were in a transition from a vegetarian to an omnivorous diet. It is theorized that extra energy from eating meat allowed brain growth, influencing the first undeniable hominid on the scene with larger cranial volumes than the East African A. Gracile.

One of them, which Richard Leakey of the National Museums of Kenya called Homo habilis, had a brain volume of about 700 cubic centimeters. Archeological evidence shows that like A. Gracile, H. Habilis also made a variety of tools. This observation supports the theory first advanced by Charles Darwin that toolmaking is both the cause and effect of walking on two legs, which frees the hands.[xxxv]

In addition, arrangements of stones at H. Habilis sites indicate that they may have constructed dwellings as well dating back more than two million years—long before the Pleistocene Ice Ages when humans sought refuge in caves.

Were we to encounter a H. Habilis today dressed in contemporary garb—we would probably give them little notice other than due to their relatively small stature. They had a high forehead like modern humans, suggesting a significant development of the neocortical areas in brain frontal and temporal lobes.

Anthropologist Ralph L. Holloway of Columbia University believes that a region of the brain known as Broca's area, one of several centers required for speech, can also be detected in H. Habilis fossil evidence. This finding suggests that the development of language, tools and culture may have occurred simultaneously.

H. Habilis, which is widely credited as the first true human, emerged during the same epoch as A. Robustus. Being larger both in body and brain weight than either the Australopithecines, the species possessed a ratio of brain to body weight about the same as that of the Gracile.

Since H. Habilis and A. Robustus emerged at the same time, it is very unlikely that one was the ancestor of the other. The A. Gracile were also contemporaries of H. Habilis but much more ancient. It is therefore possible—yet uncertain—that both H. Habilis, with a promising evolutionary future, and A. Robostus, an evolutionary dead end, arose from an earlier Gracile, Australopithecines africanus, who survived long enough to be their contemporary.

H. Habilis inhabited vast African savannahs filled with an enormous variety of predators and prey. The first modern horse appeared at about that same time. Although the horses survived, H. Habilis died out about 1.6 million years ago.[xxxvi]

In 1984, Kamoya Kimeu, a member of a team led by Richard Leakey, discovered a nearly complete 1.6-million-year-old skeleton of a young boy at Nariokotome near Lake Turkana in Kenya they named the "Turkana Boy" whose features showed considerable modern human identity. Leakey termed the species Homo erectus ("Upright Man").

Possessing endocranial brain volumes similar to ours, various other excavations sites reveal that H. Erectus had developed a sophisticated toolkit with lozenge-shaped stones, cleavers and implements. Chinese H. Erectus specimen locations which are clearly associated with the remains of cave campfires indicate that members of this group termed the "Peking Man" domesticated fire more than one half million years ago.

Throughout most of hominid history, our ancestral brain sizes, proportional to overall bodies, were approximately the same as other apes. Skull sizes of contemporary chimpanzees and Australopithecus, an extinct hominid that lived between 2.9 and 3 million years ago, averaged around 450 milliliters. By about 1.8 million years ago, hominid brains were averaging around 600 milliliters.

Paleolithic records reveal that as H. Erectus emerged with a 750 cubic centimeter brain, the neocortex—often characterized as "the rational brain"—began expanding particularly significantly. Found only in animals, this six-layered sheet-like structure in the roof of the forebrain (and particularly the outsized prefrontal cortex), enables us to hold and manipulate sophisticated concepts. Although we can't match a jaguar in a foot race, we can easily outrun all other animals in agile mental simulations.

Our modern neocortex mushroomed to its current size less than one million years ago, a remarkably short time considering our future-looking, tool-wielding, symbol-juggling human family broke off from the great apes in Africa at least six million years earlier. This human brain growth evolution was attended by neural system wiring changes that augmented processing power for functional tasks. There was a particularly significant development advancement in the Broca's area, the part of the frontal lobe that is correlated with language.[xxxvii]

Brain size growth also came with a necessity to enlarge the female pelvic girdle structure to accommodate birthing of larger-headed babies.

Present adult men and women have braincases twice

larger than the volume of recorded for Homo habilis. The incomplete closure of the modern human skull at birth, the fontanelle, is very likely an imperfect accommodation for this recent brain evolution.

A big mental capacity jump during the Middle Pleistocene period between 800,000 and 200,000 years ago doubled average hominid brain sizes. Anthropologist Rick Potts, who directs the Human Origins Program at the Smithsonian Natural History Museum, has attributed this exploded size and complexity to survival requirements imposed by intense climate change adaptation pressures involving thinking patterns, behaviors and even social strategies. Habituations that had previously worked well for those living in jungle environments became ineffective in the woodland fringe environments, on the grassland savannah or on the coasts.[xxxviii]

Beyond Hunting and Gathering

Unfamiliar environmental and unstable situations during that period may have forced Homo erectus to slow down in order to analyze and strategically plan new solutions where purely habit-based strategies no longer worked. Basic survival required a more powerful information processor with better memory, reasoning abilities and social skills needed for rapid adaption to unfamiliar environmental and resource conditions.

Our successful Sapien ancestors became adept at dealing with these challenges, while the Neanderthal, our close genetic Eurasian cousins (sharing about 90 percent of the same genes) were not. Although they too used tools, buried their dead, built fires and were bigger and stronger than our ancestors, failures to adapt to climate change ended their existence.

As their world became colder, whereas the Neanderthals retreated into shrinking forests, the more inventive Sapiens expanded their food hunting and gathering territories into growing tundra environments. And whereas the physically more robust physiques of Neanderthals fit well in the context of forest ambush-style hunting short-range, the Sapiens innovated more effective strategies. They hunted in coordinated groups; developed lighter, longer-range projectiles such as spears; and created better-adapted clothing and shelter "technologies" for the colder regions.[xxxix]

Anthropologist Robin Dunbar attributed much influence over rapid expansion in the size of the human neocortex with increased cognitive demands of cultural learning.

He reasons that as groups became larger and more closely organized during the Middle Pleistocene period between 800,000 and 200,000 years ago, it became more cognitively demanding to keep track of many complex social relationships. This imposed requirement for increased mental processing power to build better memories, representational abilities and even language.[xl],[xli]

Remarkable three-fold hominid brain growth over 2.5 million years opened rapid social evolution. Evolving language capacities and growing populations promoted swarm intelligence and tribal cooperation, with super-organisms greatly exceeding individual capabilities.

About 50,000 years ago greater Sapiens brain capacities enabled a creative explosion that continues today. The Neanderthals' extinction resulted from a failure to apply comparable cultural flexibility enabling them to move beyond experience-based strategies and tried-and-true customs. By 30,000 years ago, Sapiens had displaced Neanderthals, along with all other hominid species.

Neurosurgeon Frank Vertosick characterizes human intelligence as an emergent property from large groups of cooperative societies...an ability to store past experiences for use in future problem-solving.[xlii]

Daniel Goleman agrees, observing that "new thinking holds that our sociability has been the primary survival strategy of primate species, including our own." [xliii]

Sociability is a primary human survival strategy that engages the cooperative interactions of many minds. It enables advanced intelligence necessary to learn and teach new skills, to outwit predators and to compete with adversaries. It affords a mechanism of using abilities of innumerable brains much as if it were one collective brain.

Present day human intellect can be viewed as both a product and servant of our social life. Our improvising imagination—our early intellect - gave us the behavioral and mental scaffolding to organize and manage our experiences. This cultural capacity began to occur long before human imagination invented language and word concepts.

Although Sapiens gained a variety of versatile, inventive advantages, other early humans developed primary components of what might legitimately be characterized as "cultures" as well. For example, archeologists have discovered bones of Neanderthals with severe physical handicaps, suggesting that they were cared for by relatives.

Having said this, no living creature has ever surpassed the Sapiens' ability to adapt to immigrate and innovatively adapt to such a huge variety of radically different habitats so quickly.[xliv]

A cultural and social revolution launched Sapiens into a cognitive fast lane of unstoppable developments now referred to as "human history."

Lighting Fires of Imagination

The control of fire marked a major survival and cultural development in human evolution. Domestication of fire fundamentally influenced our Sapiens ancestors' geographic dispersal, food supply and diet, hunting weaponry and tools, art and utensils, safety from predators, societal cooperation and very possibly even their evolutionary physiology.

Evidence of widespread control of fire by anatomically modern humans dates back as a gradual process to approximately 125,000 years ago. It likely began by transporting burning brush that had been ignited by lightning for heating and cooking purposes. Archaeological evidence suggests that Sapiens later figured out how to make fire with a now drill friction device that produced ignition heat by hardwood rubbing against softwood.

Sophisticated control of fire provided a source of warmth which allowed tropical and subtropical Sapiens to survive severe climate changes and to migrate into regions that even the cold-adapted Neanderthals had been unable to inhabit.

Additionally, fire allowed the expansion of human activity into the dark and colder hours of the evening, essentially increasing the length of "daytime" for social interactions. Whereas the modern human's waking day is typically about 16 hours, most mammals are only awake about half that many hours. The nighttime use of fire also served as a means to ward off dangerous nocturnal predators.

Hominids discovered that meat dried through the use of fire could be preserved for times when scarce game and harsh environmental conditions made hunting difficult. Fire also dramatically changed what and how food was consumed.

Before the advent of fire, the hominid diet was limited mostly to plants composed of simple sugars and carbohydrates, such as seeds, flowers and fleshy fruits. Cooking made starchy and fibrous foods edible and increased the diversity of foods that were available.

Over time, this development is theorized to have contributed a skeletal change. Prior to use of fire, hominid species had large premolars needed to chew harder foods such as seeds. In response to cooking, molar teeth of Homo erectus gradually shrunk. Accordingly, hominid jaw volumes decreased as well, adapting to a variety of smaller teeth.

Cooking also killed parasites, reduced the amount of energy required for chewing and digestion, and released more nutrients from plants and meat. This particularly benefited our Cro-Magnon ancestors when the frigid Ice Age climate compelled them to live more on meat, and less on plants.

The use of fire by early humans also served as an engineering tool to modify the effectiveness of weaponry. Archeological researchers excavating in an area known as the "Spear Horizon" in Germany unearthed eight 400,000-year-old fire-hardened wooden spears along with stone tools. One of the spears was found in the pelvis of a horse. A fire-hardened lance was found in the rib cage of a straight-tusked elephant at another dig site in Lehringen, Germany.[xlv]

More recent evidence dating to approximately 164,000 years ago found that humans living in South Africa in the Middle Stone Age used fire as a heat treatment for a fine-grained rock called "silcrete." Once treated, the rocks were modified and tempered into crescent-shaped blades for arrowheads. This process was also likely used to create heat-modified tools for cutting meat of killed animals.

Fire has been used to create pottery dating back at least to evidence of 26,000-year-old high-temperature kilns and ceramic technology found at the Cro-Magnon site at Dolni Vestonice in the Czech Republic. The kilns were capable of firing clay figurines at temperatures over 400 degrees Celsius. About 2,000 fired lumps of clay were found scattered nearby.

Other 20,000-year-old pottery evidence was discovered in the Xianrendong Cave in China. However, it was during the Neolithic Age which began around 10,000 years ago that creation and use of pottery became far more widespread, a time generally associated with use in connection with early agriculture.

The discovery and the use of fire, and the sharing of the benefits, may have created a sense of sharing as a group, for example, through the cooperative participation of gathering firewood. Ongoing cultural advancements which have enabled our very survival have also made us increasingly dependent upon vast networks of cooperation.

Larry Bell

THE EMERGENCE OF CIVILIZATIONS
AND CULTURES

SOCIAL COOPERATION HAS been particularly
important for Sapiens compared with other animals because in
a cognitive development context our children are born
prematurely. This being said, since humans are born
underdeveloped, they can also be educated and socialized to a
far greater extent than any other animal.

As Yuval Noah Harari notes:

> *[Whereas] most mammals emerge from the
> womb like glazed earthenware emerging from
> a kiln—any attempt at remolding will only
> scratch or break them. Humans emerge from
> the womb like molten glass from a furnace.
> They can be spun, stretched and shaped with a
> surprising degree of freedom.*

Accordingly, lone mothers with inadequate time to forage for
food for their offspring and themselves with needy babies in

tow required constant help from other family members and neighbors.[xlvi]

About 45,000 years ago humans were predominately nomadic hunter-gatherers living on a wide-ranging omnivorous diet of wild animals and plants. This menu versatility enabled us to utilize the food resources found in diverse environments.

Vigorously aggressive and incessantly waring, these small nomadic bands weren't behaviorally adapted to life in permanent communities—first clear evidence of a settled lifestyle was found in 18,000-year-old Ice Age mammoth bone houses in the eastern part of central Europe. Long-term settlement seems also to have occurred with Natufians in the Near East between 15,000 and 11,500 years ago.

By about thirteen thousand years ago, continental glaciation had begun to slowly retreat. The Pleistocene Ice Age epoch was replaced by a warm, dry period that necessitated adjustments to far more reliable food sources. As forests spread and big game inhabiting open plains became scarcer, fishing, bird and small animal hunting and gathering became increasingly vital.

Early settlement living would have depended heavily upon naturally abundant and reliable local food sources such as hazelnuts or salmon, along with methods of long-term storage. Although wild foods still remained important in the diet, it was only about 11,000 years ago that humans began to domesticate plants and animals.[xlvii]

Events that promoted farming appear to have come about after settlements had come about and means of food storage had been invented. Subsequent storage surpluses led to trade between settlements, often utilizing river routes. Agricultural settlements dating back at least 9,600 years were established along the Euphrates River in northern Syria. Such trade moving by boats and along coastline routes promoted migration and population mixing.[xlviii]

The end of the last Ice Age during the Upper Paleolithic period marked a brief Mesolithic period which witnessed the development of new tools, especially those for woodworking and fishing. This was soon followed by a new Neolithic culture of farming, a radical social transformation which began in the Middle East about 10,000 years ago and gradually spread through Europe.

Farming also began independently and at different times in Southwest Asia, Equatorial Africa, Southeast Asia, Central America and the lowland and highland South America. With the exception of the far north, agriculture had come to predominately replace hunting-gatherer lifestyles within two millennia.

A major transition to agriculture which began between around 9,500-8,500 BC in the hill country of southeastern Turkey, western Iran and the Levant slowly spread in restricted geographical areas. Archeological records of early Middle East settlements, particularly those in the Levant, reveal that the Natufians populations (currently Israel) were then still essentially hunter-gatherers who subsisted on dozens of wild cereal species. Yet even at that early time, they had learned to build stone houses, to invent new tools such as stone scythes for harvesting wild wheat, to use stone pestles and mortars to grind it and to construct granaries to store food for future needs.[xlix]

Gradual change from hunter-gatherer to permanent settlements is evidenced by Natufian artifacts dating from about 10,300 BC to 8,500 BC. The oldest large settlement of about one hundred citizens was Jericho, circa 8,000 BC, whose merchants traded grain with Anatolia and Mesopotamia for obsidian, malachite and turquoise (named for importation from Turkey) as early as 8,300 BC.[l]

Natufian descendants gradually adopted important plant

cultivation innovations. They began to lay aside some of the wild grains they gathered from harvests to sow and replenish plantings for the next season. They discovered that they could achieve better plant results by sowing the grains deep in the ground with hoes and ploughs than by haphazardly scattering them on the surface. They learned to weed fields to guard edible plants against parasites, and conceived ways to water and fertilize them.

Rivers were first used for irrigation in Jericho. By the sixth millennium BC, irrigation innovations centered on alluvial plains of Mesopotamia between the Tigris and Euphrates Rivers yielded the most productive agricultural area ever witnessed. This capacity to feed growing populations established locations for some of the world's earliest large cities.[li]

As cereal cultivation demanded effort which left less time to gather and hunt wild species, foragers transitioned to become farmers. Those tribes that resisted faced a serious dilemma.

Once agricultural settlements secured control, demographic growth in the area enabled farmers to overcome hostile challengers in small foraging bands by sheer weight of numbers. Their only remaining choices were to run away, abandoning their hunting grounds to field and pasture, or to take up the ploughshare themselves. In either case, their old lifestyle was forever over.

The first domesticated forms of Einkorn, a wild cereal, were developed about 10,500 years ago. Modern wheat arose about 7,000 years ago in Iran from a cross between a wild grass and a more recent emmer wheat domesticated in the Euphrates valley of modern Syria.[lii]

Whereas 10 thousand years ago wheat was merely one wild grass among many, within a few short millennia it was growing all over the world. Wheat fields are now estimated to cover about 870,000 square miles of the globe's surface—nearly ten times the size of Britain.[liii]

More than 90 percent of the calories that feed humans today came from a handful of plants that our ancestors domesticated between 9,500 BC and 3,500 BC. First were wheat, rice, barley potatoes and maize (called corn in the United States); followed by peas, lentils and potatoes around 8,500 BC; olive trees about 5,000 BC; and grapevines in 3,500 BC. Cashew nuts were domesticated somewhat later.

Concurrently, a new agricultural revolution progressed with the domestication of plants; our Sapiens ancestors also tailored to animals to serve desired purposes.

Dogs became human-domesticated partners about 15,000 years ago, likely supporting transitions from foraging to settled societies in a variety of ways. Canines could be trained to hunt, and their alertness to humans and animals sounded barking sentry warnings. They served as welcome cold night bed warmers, and also provided a meat source in emergencies.

DNA records indicate that dogs were first bred from gray wolves which then rapidly spread throughout Eurasia. The earliest dog fossil dates back to 14,000 years ago in Germany.

Domesticated sheep, goats, cattle (from aurochs—an extinct breed of European cattle) and pigs (from wild boar) appeared between about 10,000 and 9,500 years ago. Horses were first domesticated on the Eurasian steppes around 6,000 years ago.[liv]

The evolution of farming cultures introduced the domestication of animals in accompanying stages. Broadly summarized, this process likely began as nomadic Sapiens hunters first altered the constitutions of the herds through a process of selective hunting of older, sicker animals, sparing fertile females and youngsters to safeguard the long-term vitality of their food supply.

A second step may have been to actively defend herds from predators such as lions, wolves and even rival human bands. This would have included driving interlopers away, as well as corralling herds into a narrow gorge in order to better control and defend it.

The final big domestication stage involved a more careful selection among members of the herds in order to tailor them to preferred features. Those individuals that exhibited the most aggressive resistance to domestication control—along with those with skinniest meat content—were slaughtered first.

Included by selective features were horses, cattle, donkeys, sheep, boars and chickens which provided food (meat, milk and eggs), raw materials (skins, wool) and muscle power (transportation, ploughing, grinding and other physical tasks).

Animal domestication led to special new herding and agriculture societies needed to feed ever-growing populations. Since then, the few million sheep, cattle, goats, boars and chickens that existed in Afro-Asian niches ten thousand years ago has grown dramatically. Today's world contains about a billion sheep, a billion pigs, more than a billion cattle and more than 25 billion chickens...the most widespread fowl ever.

Early domestic farming experiments often led to big setbacks. Among these, diseases were inordinately potent consequences. Malaria is thought to have become common among humans 10,000 to 5,000 years ago in connection with West African introduction of slash-and-burn agricultural practices which left sunlit pools as breeding sites for Anopheles mosquitos. Other diseases transferred from livestock to people promoted infectious epidemics.

Archaeological bone and tooth evidence reveal that many of these early farming efforts also led to health problems not experienced by hunter-gatherers. Increased food production often couldn't keep pace with population growth. In particular, a lack of knowledge regarding fertilization and soil depletion degraded crop and livestock yields.[lv]

Advancing settlement and farming lifestyles presented major challenges requiring new ways of thinking and social relationships. Settlement living promoted an increasing need for social cooperation commensurate with population densities. Founded on a principle of reciprocity, such cooperation not only served to enhance community prosperity and stability, but also promoted a social form of competitive natural selection that drove individuals and collaborating groups to achieve higher accomplishments.

Less fortunately, that increased social interdependence also gave rise to a new social problem whereby "freeloaders" and other cheats deceived others to further their own interests. Although human traits of cunning and deceit had long-existing history predating settlement cultures, close living and work-sharing circumstances required effective means to apply peer pressure—sometimes with very painful or mortal infraction penalties.

Whereas kinship was the primary tie among hunter-gatherer groups, increasing population densities required trading nomadic independence and equality for collaborative benefits of conformity within allegiance to governing domain rules and community hierarchy status.

Private property became an innovation that established class by different levels of ownership. Chiefs, commoners, rich and poor families, labor specialization and fixed social rules of behavior initiated life complexities that have continued to increase over time.[lvi]

The Revolutionary Human Species

Our fortunate ancestral cognitive and social inheritance has led to even faster-paced innovations. Imagine that while 11 millennia passed between the Agricultural Revolution and the Industrial Revolution, it only took only 120 to get from the Industrial Revolution to the light bulb. Humans landed on the Moon 90 years later. Twenty-two years after that came the World Wide Web. A mere nine years later, the human genome was fully sequenced.[lvii]

Archaeological records reveal few changes in archaic human social patterns or new technologies prior to a Cognitive Revolution which began about 70,000 years ago. Following the emergence of Homo erectus, stone tools which are broadly recognized as a defining feature of our species remained roughly the same over nearly 2 million years.

The emergence of Homo sapiens was a huge cognitive and behavioral game-changer—one requiring no significant new genetic or environmental dependencies. As noted by Yuval Noah Harari:

> *Consequently, ever since the Cognitive Revolution, Homo sapiens has been able to revise its behavior rapidly in accordance with changing needs. This opened a fast lane of cultural evolution, bypassing the traffic jam of genetic evolution. Speeding down this fast lane, Homo sapiens soon far outstripped all other human and animal species to cooperate.*[lviii]

Harari characterizes the Cognitive Revolution as the point when history declared its independence from biology. Until then, "the doings of all human species belonged to the realm of biology, or if you so prefer, prehistory." (Harari clarifies that he prefers to avoid the term "prehistory," because it wrongly implies that even before the Cognitive Revolution, humans were in a category of their own.)

Whereas a hike in East Africa two million years ago would likely have revealed what appeared to be a familiar cast of human characters, Harari points out that their behaviors were not really so different than many other animals. Just as archaic humans loved, played, formed close friendships and competed for status and power—so did chimpanzees, baboons and elephants.

The period from about 70,000 years ago to about 30,000 years ago witnessed the invention of boats, oil lamps, bows and arrows and needles (essential for sewing warm clothing). Then within a remarkably short period, about 45,000 years ago, Sapiens had reached Europe, East Asia and had somehow crossed the open sea to Australia—a continent previously untouched by humans.

Sapiens, whose bodies were originally adapted to living in the African savannah, devised ingenious solutions which enabled them to migrate to brutally frigid climates such as northern Siberia which even cold-adapted Neanderthals avoided. They learned to make snowshoes and effective thermal clothing composed of layers of furs and skins, sewn together tightly with the help of needles.

As their thermal clothing and hunting techniques improved, sapiens dared to venture deeper and deeper into the frozen regions. More advanced weapons and sophisticated hunting techniques enabled Sapiens to track and kill mammoths and the other big game of the far north. Every mammoth was a vast quantity of meat (which, given the cold temperatures, could be frozen for later use), tasty fat, warm fur and valuable ivory.

Whereas early Sapiens inhabited existing caves or rock shelters where available, only very recently, especially within the last 20,000 years, were those natural shelters enhanced with interior walls or other simple modifications.

Shelters in open areas were often constructed using a range of framework materials including wooden poles and bones of large animals, such as mammoths. These structures were probably covered with animal hides. Habitat sites sometimes included fire hearths.

Around 14,000 BC, the Sapiens chase took some of them from north-eastern Siberia to Alaska. At first, glaciers blocked the way from Alaska to the rest of America, allowing no more than perhaps a few isolated pioneers to investigate lands further south. However, around 12,000 BC, global warming melted the ice and opened an easier passage.[lix]

Making use of the new Bering Straits corridor, our ancestors moved south en masse, spreading over the entire North American continent. And although originally adapted to hunting large game in the Arctic, they soon adjusted to an amazing variety of new climates and ecosystems.

Within merely a millennium or two, Siberian descendants settled the thick forests of what became the eastern United States, the swamps of the Mississippi Delta, the deserts of Mexico and steaming jungles of Central America. Some made their homes in the river world of the Amazon basin, while still others struck roots in Andean mountain valleys and open pampas of Argentina.

Shortly after man entered North America in the Pleistocene period, there were massive and spectacular kills of large game animals, often by driving them over cliffs. This ability to stalk a single wildebeest or to stampede a herd of antelope to their deaths indicates that the hunters shared at least some minimal symbolic verbal language.

While skeletons of H. Erectus suggest the absence of muscle control for respiratory speech recognition, some rudimentary form of vocalization probably existed as much as

2.5 million years ago. Positioned much lower in the throat than for apes, the upright posture of Erectus favored the evolution of a larynx enabling enunciation of consonant and vowel sounds. Also, the upright bipedal breathing posture facilitated specialized muscle development for high-fidelity sounds.

Darwin pointed out that gestural languages cannot usefully be employed while our hands are otherwise occupied, or at night, or when our view of the hands is obstructed. One can then imagine gestural languages being gradually supplemented and then supplanted by verbal languages—which may have originally imitated the sound of the object or action being communicated.

Although any language ability we inherited from H. Erectus, the first Homo species to hunt as a major means of subsistence, would have been very limited, it would still have been superior to that of H. Habilis. In any case, they managed to make sophisticated stone tools, and to extend their range beyond Africa.

Oldwan stone implements, including simple choppers and pounders, continued with little innovation for about a million years between 2.6 and 1.8 million years ago. Subsequent tool-making industries, such as the Acheulean (1.7-0.2 MYA) and Mousterian (30,000-40,000 MYA), were more innovative and diverse.

Production of simple hand axes that were flaked repeatedly to create a biface symmetrical point remained basically the same for over a million years before evolving to create other adaptive tools like spear points, arrowheads and stone knives.

Here, what is typically characterized as the "stone age" might more accurately be described as the "wood age." Since archeological evidence consists mainly of bones and stone tools, artifacts made these ancient hunter-gatherers of more perishable materials such as wood, bamboo and leather would leave fewer fossil traces.

Yet soon after emerging in the last Ice Age more than 30,000 years ago, Cro-Magnon Sapiens in southern France invented elaborate tools, spear throwers, fully barbed harpoons and flint master tools for making hunting weapons.

These early humans learned how to secure handles to stone for hand axes with tree resin or bitumen, using a lever principle for increased power and velocity for work and distance killing. They innovated arrows with wooden shafts. They stitched together animal skins with bone needles for caps, shirts, jackets, trousers and moccasins. They built mammoth bone houses, laid stone floors, made animal fat burning lamps and they built fireplaces for heating and cooking.[lx]

Visual technologies soon advanced and spread as well. Cave paintings, such as the Chauvet Cave in the Gorges de l'Ardèche, France dating back 30,000 years ago, demonstrate that human minds could convert three-dimensional animals (e.g., a bison or bear) into two-dimensional line representations.[lxi]

Although Sapiens had relatively very simple cultures compared to now, they were far more advanced than any previous species. As they evolved and expanded, our ancestors brought a trading culture with them. Included were social prestige items such as shells, amber and pigments.

Archaeologists excavating 30,000-year-old Sapiens sites in the European heartland have discovered seashells from the Mediterranean and Atlantic coasts. In all likelihood, these shells got to the continental interior through long-distance exchanges between different Sapiens bands. Neanderthal sites, on the other hand, lack any evidence of such commerce.[lxii]

Humans are by no means unique among other animals in forming social bonds and relationships; ours are by far most complex. However, we are the only creatures that can truly make any sense out of trading something truly needed—food to eat, or weapons to hunt with, for example—for a bauble of adornment or coin of currency with no practical intrinsic value.

As Yuval Noah Harari notes, a unique Sapiens trait is our ability to create and believe in our own fictions to create imagined realities. All other animals use their communication system to describe reality. While not all fictions are shared by all humans, at least one—money—has become universal. Harari observes that while dollar bills have absolutely no value except in our collective imagination, everybody believes in them.[lxiii]

Such bonds and collaborations between Sapiens provided a crucial edge over other human species. Relations between neighboring tribes eventually became tight enough to merge into single tribes that shared common languages, common myths, common norms and common values. These combined tribes exchanged members, hunted together, traded luxuries, cemented political alliances, celebrated religious festivals and honored sacred rituals.

Humans are the only animals that both anticipate their mortality and contemplates what may follow. Burial ceremonies that include the interment of food and artifacts along with the deceased go back at least to our Neanderthal cousins, suggesting not only a widespread awareness of death, but also an already developed ritual ceremony to sustain the deceased in the afterlife.

Burials were infrequent and very simple prior to 40,000 years ago. They later began to become much more elaborate with the inclusion of valued objects such as tools and body adornments. Red ochre was sprinkled over many of the bodies prior to burial.

In 1955, archaeologists discovered the skeleton of a fifty-year-old man covered with strings containing a total of 3,000 mammoth ivory beads in a 30,000-year-old burial site in Sunghir, Russia. This adornment, together with 25 mammoth ivory wrist bracelets and a hat covered with fox teeth indicated that he was probably an important leader of a hierarchical society.

Archaeologists also discovered another Sunghir tomb containing two skeletons buried head-to-head. One belonged to a boy aged about twelve or thirteen, and the other a girl about nine or ten. The boy who was covered with 5,000 ivory beads wore a hat and a belt containing a total of 250 fox teeth (at least sixty dead foxes). The girl was adorned with 5,250 ivory beads. Both children were surrounded by statuettes and various ivory objects.

One conclusion suggests that both were children of very important tribal figures. An alternate theory is that they may have been ritually sacrificed—possibly as part of ceremonial burial rites of a leader.[lxiv]

Not all encounters between Sapiens bands were friendly. Some groups violently fought one another over resources and for other unknown reasons.

A 12,000-year-old cemetery in the Sudan evidenced that 24 of 59 skeletons discovered there were found with arrowheads and spear points embedded in them or lying nearby. The skeleton of one woman revealed 12 injuries.

Archaeologists at Ofnet Cave in Bavaria discovered the remains of 38 foragers, mainly women and children, who had been thrown into two burial pits. Half of the skeletons, including those of children and babies, bore clear signs of damage by human weapons such as clubs and knives. Those few belonging to mature males bore worst marks of violence. It is suspected that an entire forager band had been massacred.[lxv]

Larry Bell

Powers of Communication and Conceptualization

Referring to the Sapiens advantage over Neanderthals he attributed to "cognitive fluidity," author and University of Reading Deputy Vice Chancellor Steven Mithen noted that Sapiens gained a mental capacity which led to an explosion some 50,000 years ago in language, technology and art. Mithen pointed out that whereas Neanderthals lacked the "ability for metaphor and had limited imagination," cognitive fluid thought requires an ability to make connections between modular thinking domains which involve perceptions of social, material and natural worlds.[lxvi]

Stephen Asma cites the development of language as both a cognitive product and enablement of Sapiens' rapid evolutionary success. He writes:

> Most evolutionary psychologists claim that the cause of [the Homo sapiens'] cognitive fluidity was the development of language (in the late Pleistocene), because language provides an obvious syntactical/ grammatical system for manipulating representations.[lxvii]

Asma contemplates:

> I try, for example, to imagine what it was like to be a conscious being before language (either a Homo erectus man or a contemporary Homo sapiens baby), I run straight into the fact that my mind is already deeply structured by language. It is difficult to peek around the veil of language to see the pre-linguistic operating

68

system at work.

Eric Kandel literally visualizes an answer to Asma's dilemma. In his book *The Age of Insight,* he writes:

> *Perhaps in human evolution the ability to express ourselves in art—in pictorial language - preceded the ability to express ourselves in spoken language.*[lxviii]

Lacking true language, and prior to pictorial images, our pre-Sapiens ancestors in Africa and Eurasia around 500,000 years ago probably communicated with one another by gesture and mimicry. This view is consistent with anthropologist records revealing that bands of hunter-gatherers communicated by means of gestures, facial expressions and mime.[lxix]

In contrast with toolmaking, innovation in the human visual art tradition exploded in a very short and recent period. The earliest forms of pre-figurative decorative design reach back to about 140,000 years ago.[lxx]

Although rare evidence of symbolic behavior can be traced back to a number of African sites about 100,000 years ago, these artistic expressions appear more as flickers of creativity with little evidence of sustained expression. It is not until about 40,000 years ago that complex and highly innovative cultures appear and include behavior that would be recognized as typical of modern humans today.

Many researchers believe that this explosion of artistic material is at least partly attributable to a change in human cognition—an ability to think and communicate symbolically or to memorize better. Some also theorize that expanding tribal sizes and more complex social structures played key roles.

Dating back about 35,000 to 40,000 years ago, the Upper

Paleolithic Löwenmensch mammoth ivory figurine or "Lion-man" of the German Hohlenstein-Stadel cave is the oldest-known uncontested example of figurative art. Carved gouges indicate that it was carved using a flint stone knife.

Stone-sculpted Venus figurines which appeared across Europe, from Portugal to Russia, between 28,000 and 21,000 years ago all had a remarkably similar style with exaggerated breasts. They are likely to reflect culturally-connected fertility symbols or mother goddess references.[lxxi],[lxxii]

Although decorative marine shell beads found in Israel date back about 90,000 years ago, and ostrich shell beads found in Morocco date back about 80,000 years, items of personal adornment (which were not sewn into clothing) only became prolific about 35,000 years ago. This suggests that human culture had come to attach growing importance to visual symbols of social appearance and status.

Musical instruments also began to appear around this time. Decorated mammoth bones shaped into castanets and flutes were found at various French Paleolithic Cro-Magnon sites ranging from 30,000 to 10,000 years old.[lxxiii]

Ever since the beginning of the cognitive revolution about 70,000 years ago, Sapiens have thus been living in a dual reality. One reality is directly observable and experienced, an objective reality of rivers, trees and lions. The other is an imagined reality, one of gods, laws and tribal customs.[lxxiv]

Visual arts that soon followed reveal an apparent merging of real-life story-telling with afterlife imaginings. Paleolithic paintings dating about 15,000-20,000 years ago in the Lascaux Cave in the Dordogne region of southwestern France depict what appears to be a man with the head of a bird and erect penis being killed by a bison. Another bird beneath the figure might possibly be interpreted to symbolize the soul released from the body at the moment of death.

What Makes Humans Truly Exceptional?

Wall paintings Las Cueva de las Manos ("Cave of the Hands") in the valley of the Pinturas River, in Argentina, date back roughly 10,000 years. Although there are three distinct styles which are believed to have been created at different time periods, the highlight is the hundreds of colorful handprints dated to around 5,000 BC.

It's believed these cave dwellers stenciled their own hands using bone-made pipes to create the silhouettes of their left hands, indicating that they probably held the spraying pipe in their right hands. Various mineral pigments were used to make different colors—iron oxides for red and purple, kaolin for white, natrojarosite for yellow and manganese oxide for black.

The cave also contains hunting scenes and representations of animals and human life dating back further to around 7300 BC. The hunter-gatherers who lived in the caves depicted pursuit of prey including the guanaco, a type of llama using the bola—cords with weights on either end which are thrown to trap the legs of the animal. A third category of paintings depicted animals and humans in a more stylized and minimalist fashion, done largely in red pigments.

Cognitive fluidity which emerged during the Upper Paleolithic period roughly 50,000 years ago enabled our ancestors to advance tool technologies, cooperative game hunting, tribal societies and ceremonies, language and art—an explosion of ever-accelerating creative progress which now grows at an exponential rate.

Language, including visual art, led to innovative story-telling lessons and mythologies which established and passed on post-generational societal mores, governance and status hierarchies, traditions and rituals and behavioral expectations and enforcements.

Written Languages for Cooperation and Conflict

Conflicts arising from continuous, close contact and community defenses against marauding hunter-gatherers promoted the development of more and more elaborate writing communication methods essential for negotiation and collaboration.

Long-distance trade in Mesopotamia created a need to be able to communicate across the expanses between cities or regions. Written records used exclusively for such accounting purposes were a variety of more than 500 cuneiform or wedge-shaped pictogram symbols pressed into clay which served to aid in keeping track of such things as which parcels of grain had gone to which destination or how many sheep were needed for events like sacrifices in the temples. Many of these surviving records had to do with sales of beer, a popular beverage.

A type of phonetic writing which appeared after 2,600 BC was characterized by a complex combination of word-signs and phonogram—signs for vowels and syllables—that allowed the scribe to express ideas. Impressed on clay tablets, they were applied for a vast array of economic, religious, political, literary and scholarly documents.

The first alphabet which emerged in Mesopotamia around 2,500 BC was confined to consonants. The Greeks later created vowels the middle of the 8th century BC, completing an alphabet which continues today with only minor changes.

The first writer in history known by name is the Mesopotamian priestess Enheduanna, daughter of Sargon of Akkad, who wrote poems dedicated to the goddess Inanna.

Predating Homer and the historical part of the Bible, the first recording of major literature dates to 2,100 BC. Gilgamesh tells a saga of heroic King Ukruk and a flood which decimated

Sumerian cities which has striking similarities to the Great Flood of Noah legend which was passed down to Assyrians and Hebrews. Evidence that such a flood occurred consists of a ten-foot deposit of silt at the location.[lxxv]

Between 2,500 and 1,300 BC, Aryan warriors conquered most of India, merging their Semitic alphabet with Sanskrit of India to create new literary Brahmi script. Ancient Rig Veda Hindu poems and songs reveal a scarcely literate, egalitarian society prior to its takeover by a militaristic, patriarchic alphabetical one.[lxxvi]

Hindu Brahmins established strict laws forbidding contact with writing, and consequently, the Hindu civilization existed for two thousand years without a written law. Laws of Manu were later established following adoption of the Brahmi script.

Inventions of farming, language, the alphabet and writing catalyzed transformational new social arrangements, accountabilities, allegiances and alliances.

Writing's important contributions to these developments were many. Writing codified understandings between individuals, rules of behavior and penalties for noncompliance.

Writing provided a medium to document, disseminate and immortalize events and ideas for then-present and future generations (including us).

And, in various cases, both for better and worse, writing served as an instrument of power which was applied and exploited by social hierarchies and anointed leadership elites to exert dominant control. Insidious catch-22 consequences often resulted in the co-existence of cultural stability and advancement on one hand, and the concomitant promulgation of social exploitation, oppression and warfare on the other.

The Rise of Empires and Societies

Prior to the agricultural revolution, those few human enclaves that existed were very small, surrounded by expanses of untamed nature. By the first century AD, only 1-2 million of the previous 5-8 million nomadic foragers still remained (mainly in Australia, America and Africa). Their numbers had become dwarfed by the world's 250 million farmers.[lxxvii]

The rapid expansion of permanent farming settlements along with trade and cooperative regional defense relationships created a need for more elaborate and absolute governance means to coordinate and enforce populace behaviors deemed mutually beneficial. Who actually deemed what was most beneficial, and beneficial to whom, varied greatly from one region to another, with great consequences to those governed or resisting sovereign authority.

A capability to feed a thousand or more people in the same town or region offers no assurances that individual members and groups within that population will easily agree how best to divide the land and water, how to settle disputes and conflicts and how to act together in times of drought or war. Common rules and penalties must also be established and enforced to ensure cooperative compliance.

The first millennium BC witnessed the appearance of three primary unifying "orders" of law and governance:

- The first, an economic order, perceived the known world as a single market, and all populations as customers.
- The second universal order was an imperial order, where conquerors saw the world their empire and all populations as their prospective subjects.
- The third universal order was religious, where prophets and believers viewed the entire world bound by particular articles of faith applicable to everyone everywhere such as Christianity and Islam.

Whereas emerging cities and regional domains such as kingdoms began to cooperate with certain others outside, who they called "brothers" or "friends," there was always an inevitable natural tendency to discriminate between the "us" versus "them"—those less worthy "barbarians" who resided in the next valley or beyond the far mountain range.

Deadly conflicts sometimes arose when those outsiders possessed fertile land and material wealth coveted by insiders or their leaders. A popular solution was to establish more powerful empires needed to conquer them and either make them part of "us"– or alternatively, to either enslave or exterminate "them." This occurred, for example, when Romans who invaded Scotland in 83 AD encountered fierce resistance from local Caledonian tribes reacted by laying waste to the country.

And sometimes populations who actually were conquered eventually ceased to view the empire as an alien system of occupation whereby the ruled and rulers alike viewed the other "them" as a new "us." After centuries of imperial rule, all Roman subjects were granted citizenship. Some non-Romans even rose to top ranks in the officer corps of the Roman legions.

Powerful empires need not necessarily emerge from military conquest. The Athenian Empire began as a voluntary league, and the Hapsburg Empire came about through a string of mutually beneficial marriage alliances.

Nor must an empire be ruled by autocratic leaders. The empires of Novgorod, Rome, Carthage and Athens were governed as democracies, as was the enormous British Empire of modern times.

Size doesn't always determine empire status either. The Athenian Empire at its zenith was much smaller in size and population than today's Greece, and the Aztec Empire was smaller than today's Mexico.[lxxviii]

Fundamentally, an empire is a political order with two important defining characteristics. First, it must rule over a significant number of distinct peoples, each possessing a different cultural identity and a separate territory.

Second, empires are characterized by flexible borders and an ability to swallow and digest more and more additional nations and territories without altering their basic structure or identity. Together, cultural diversity and territorial flexibility gives empires not only their unique character, but also their central role in history.

Some empires were clearly much mightier than others. Some last for thousands of years, while others last only months. Some stretched across multiple continents, while others only existed on one single continent. And some empires were more savage than others, depending upon numerous wars and brutal punishments to maintain their power structures.

The inventive and enterprising Sumerians of ancient Mesopotamia who built canals, dams and reservoirs to support Mesopotamian agriculture and human sustainment before 4,000 BC also created violent empires headed by cunning rulers wielding absolute power.

During the fifth and fourth millennia BC, cities with tens of thousands of inhabitants sprouted in the Fertile Crescent, a quarter-moon-shaped region from the Persian Gulf through modern-day southern Iraq, Syria, Lebanon, Jordan, Israel and northern Egypt. Each of these cities held sway over many nearby villages. The rapid growth of city-states incited a mad scramble for wealth. Empires came into being in response to boundary disputes over irrigated lands and control of valuable resources including metals, stone and timber.

Among multiple conflicts, the first war in 2,525 between

Umma and Lagash, two cities nearly 18 miles apart, was over the fertile region of Guendena. Remnants of memorials commemorate the victory over the king of Umma at the hands of Eannatum of Lagash. Such defeats typically ended badly for captured opponents who were carted away as slaves.

Around 2250 BC, Sargon the Great forged the first true empire, the Akkadian, with over a million subjects and a standing army of 5,400 soldiers. Sargon began his career as a gardener, and then as King of Kish, a small city state in Mesopotamia. Within a few decades he managed to conquer not only all other Mesopotamian city-states, but also captured large territories outside the Mesopotamian heartland and covered much of modern-day Iraq.

After the Akkadian empire fell, it was succeeded by two great mega-empires with millions of soldiers—Babylonia to the south, and Assyria to the north. Assyrians were known to threaten their enemies and captives with savage brutality. Two thousand years later when the Assyrians were defeated, its nobles were massacred in revenge for their tyranny.

In 3100 BC, the entire lower Nile Valley was already united into an Egyptian kingdom, where pharaohs came to rule an empire covering thousands of square miles and ruling hundreds of thousands of people. Unlike other major empires, Egypt remained stable without warfare for nearly three thousand years. This stasis ended with the Roman conquest of Egypt in 30 BC.

Surrounded on three sides by deserts, the Isthmus of Suez and the Mediterranean on the remaining side, Egypt's borders had previously required little defense. Consequently, when the first pharaoh, Menes, united Egypt around 3,000 BC, the alien "barbarians" beyond those natural borders presented little interest other than to the extent that they had land or natural resources that the pharaoh wanted.

Avoiding great conflicts with neighboring empires and supported with fertile agriculture, the pharaohs turned priority attention, power and wealth in preparing for their own afterlives. Historian Charles Van Doren theorizes that this may explain why other than great feats of engineering exemplified by the Great Pyramids, Egypt's technological progress advanced less rapidly than in many other empires. In other words, invention wasn't essential for survival or better living.[lxxix]

In 221 BC, the Qin dynasty united China, where taxes levied on 40 million Qin subjects paid for a standing army of hundreds of thousands of soldiers, and a complex bureaucracy that employed more than 100,000 officials. Shortly afterwards, Rome united the Mediterranean basin.

The Mongol Empire which existed during the 13th and 14th centuries emerged from the unification of several nomadic tribes under the original leadership of Genghis Khan. The largest contiguous land empire in history, its territory eventually stretched from Eastern Europe and parts of Central Europe to the Sea of Japan, extending northwards to Siberia, eastwards and southwards into the Indian subcontinent, Indochina and the Iranian Plateau, and westward as far as the Levant and the Carpathian Mountains. Connecting the East and the West, it allowed the dissemination and exchange of trade, technologies and ideologies across Eurasia.

Any resistance to Mongol rule was met with massive collective punishment. Before conquering a city, they would send a messenger to demand surrender. No harm would come to those who chose to surrender, but unbelievable brutality would meet those who did not.

In 1258 AD when Baghdad refused to surrender to Genghis Khan's grandson Hulagu Khan, its population was massacred, the city was sacked, and its libraries of rich information were destroyed. As a special warning to others who might consider defying Mongol orders, the city ruler was wrapped up in a rug and beaten to death.

The Mongol populace was governed by a code of law called "Yassa," meaning "order" or "decree," with strictly enforced canons carrying severe penalties. For example, it decreed a death penalty if one mounted soldier following another did not pick up something dropped from the mount ahead of it, and also meted out harsh responses to rape and murder.

The Yassa guaranteed its subjects some important freedoms as well. Chiefs and generals were selected based upon merit, those of rank shared much of the same hardships of the common man. Virtually every religion, including Buddhism, Christianity, Manichaeism and Islam were assured freedom to be practiced, and all religious leaders were exempt from taxation and public service.

By 1450 AD, close to 90 percent of all humans lived in a single mega-world: the world of Afro-Asia. Most of Europe, and most of Africa (including substantial chunks of sub-Saharan Africa) were then already connected by significant cultural, political and economic ties.

Most of the remaining tenth of the world's population was divided between four worlds of considerable size and complexity:

- The Mesoamerican World, which encompassed most of Central America and parts of North America.
- The Andean World, which encompassed most of South America.
- The Australian World, which encompassed the continent of Australia.

- The Oceanic World, which encompassed most of the islands of the south-western Pacific Ocean, from Hawaii to New Zealand.

Over the next 300 years, the Afro-Asian giant swallowed up almost all the other worlds. It consumed the Mesoamerican World in 1521, when the Spanish conquered the Aztec Empire. At about that same time it began to overtake the Oceanic World during Ferdinand Magellan's circumnavigation of the globe, then soon afterwards completed that conquest. Spanish conquistadors advanced to crush the Inca Empire in 1532, collapsing the Andean World.

After the first European landed on the Australian continent in 1606, British colonialism began in earnest in 1788. Fifteen years later the Britons established their first settlement in Tasmania, thus bringing that last autonomous human world into the Afro-Asian sphere of influence.

Spread of Religious Orders and Disorders

Religion has constituted both a great unifying (and dividing) influence of humankind. Whereas it has served as the glue that binds social stability to deal with common threats, it has also functioned as a cohesive medium to secure rule by tyrants.

Ruling submissive subjects through violence and cunning, some sovereigns asserted their own personal supreme divinity status. Others enlisted alliances with prophets and priests to impose strictures of despotic religious authority.[lxxx]

Broadly defined, the term "religion" pertains to a system of human norms and values that is founded upon a belief in some form of superhuman order. Further, a religion typically asserts which particular norms and values are to be considered most important and binding as ordained by an absolute and supreme authority.

What might be described as a "religious revolution" might trace back to the transition from hunters-gatherers to farmers at the early beginnings of the agricultural revolution.

Any sense of supremely ordained human status might not have readily occurred to those early ancestors who spent their lives picking wild plants and pursuing wild animals they regarded to be natural equals. To them, the fact that they hunted sheep did not make sheep inferior to them, just as the fact that tigers hunted them did not make them inferior to tigers.

This likely began to change after people domesticated some of those sheep and became responsible for keeping their now-human-dependent flocks safe from other human and animal predators and healthy from disease epidemics arising from closer confinement.

The help of deities such as a fertility goddess, sky god and god of medicine might first have been called upon to mediate between humans and their newly co-dependent plants and animals. Religious liturgy for thousands of years after the agricultural revolution focused heavily on sacrifices of lambs, wine and cakes to divine powers in exchange for fecund flocks and abundant harvests.

Chronic, often bloody, ideological conflicts later arose regarding whether one particular god or a more specialized power-sharing group of gods wielded divine supremacy. Those subscribing to a single god supremacy (or theocracy) are termed monotheists. Other faiths that understand the world to be controlled by a group of powerful gods are known as polytheistic religions (from the Greek "poly" = many, "theos = god").

The first known monotheist religion appeared in Egypt around 1,350 BC when Pharaoh Akhenaten declared that the god Aten (but one of several minor deities of the Egyptian pantheon) wielded supreme power over the Universe. Worship of Aten over all other gods became institutionalized as the state religion. Akhenaten's religious decree ended with his death when the worship of Aten was abandoned in favor of the old pantheon.

Monotheists have tended to be far more fanatical and missionary than polytheists. This behavior arises from a prevalent belief that since they are in possession of the entire message of the one and only true God, they are therefore compelled to discredit all other religions. This is evidenced by countless violent wars over the last two millennia as monotheists have repeatedly tried to exterminate all religious competition.[lxxxi]

As noted by Yuval Noah Harari, the monotheists are clearly winning. At the beginning of the first century AD, there were hardly any monotheists in the world.

Throughout its history, Judaism has not been a missionary religion. Yet through missionary zeal, a small Jewish sect took over polity of the mighty Roman Empire and established a spiritual Catholic headquarters in the capital city. Within the next 500 years, missionaries were busy spreading Christianity to other parts of Europe, Asia and Africa.

Rapid Christian expansion served as a precedent for another monotheist religion that appeared in the Arabian Peninsula in the seventh century—Islam. Like Christianity, Islam began as a small sect led by a charismatic prophet, spread throughout vast lands, and came and continues to play a central role in world history.

Born in Mecca around 570 AD, Muhammad, the Prophet of Islam, was orphaned at age six, raised by his uncle Abu Talib. Working as a trade merchant between the Indian Ocean and the Mediterranean Sea, he became greatly respected for trustworthiness.

At the age of about 40, Muhammad began to experience visions and repeatedly sought solitude in a cave on Mount Hira on the outskirts of Mecca. It was there that an angel, Jebreel (Gabriel), appeared to him and revealed messages from the one true God, Allah, which are recorded in the Quran.

In the year 610, Muhammad began preaching publicly, proclaiming that "God is One." Although he slowly won over a small group of ardent followers, including his wife, repeated efforts to attract Jews to his cause were unsuccessful.

Muhammad's monotheistic teachings which impugned the traditional polytheistic worship of gods and goddesses provoked powerful merchants and other Meccan city leaders to persecute him. He and his small band of followers were ultimately forced out of the city in 622 AD. The group established new residence in Medina (then known as Yathrib) where Muhammad united tribes under a "Constitution of Medina" and gathered together an army of about 10,000 converts.

In 628, Muhammad negotiated a truce with the Meccans, and in the following year, returned as a pilgrim to the city's holy sites. However, the murder of one of his followers provoked him to march upon the city, which soon surrendered. There was little bloodshed, with Muhammad only demanding of the Meccans that pagan idols be destroyed.

Muslims continue to revere Muhammad as the final prophet of God and the words of his messages recorded in the Quran are regarded as ultimate truths.

The world of Islam rapidly expanded as Egyptians, Syrians and Mesopotamians came to increasingly be dominated by non-Arab Muslims, in particular by Iranians, Turks and Berbers. Already by the time of Muhammad's death in 632, most of the Arabian Peninsula had been converted to the religion.

By the end of the first millennium AD, most people in Europe, West Asia and North America were monotheists, and empires from the Atlantic Ocean to the Himalayas claimed to be ordained by the single great God.

By the early sixteenth century, monotheism dominated most of Afro-Asia, with the exception of East Asia and the southern parts of Africa. It then began to gain dominance moving towards South Africa, America and Oceania. Most people outside East Asia now adhere to one monotheist religion or another, and the entire global political order is built on monotheist foundations.[lxxxii]

Nevertheless, many polytheists also essentially subscribe to a belief in a supreme God power, albeit one that presides over all lesser gods. In classical Greek polytheism, for example, Zeus, Hera, Apollo and their colleagues were subject to an omnipotent and all-encompassing power—"Fate" (Moria, Ananke).

The Hindus built no temples to their supreme God, Atman, for the same reason that the Greeks built none to Fate. Rather than seeking assistance from Atman in addressing such local worldly matters as winning wars and healing illnesses, these fall more into the domain of more specialized deities specialized such as Ganesha, Lakshmi and Saraswati.

Here, Fate and Atman are too busy dealing with big Universe issues to be concerned with mundane desires, cares and worries of humans. From such an all-encompassing spiritual vantage point, it makes no difference whether a particular kingdom wins or loses, whether a particular city prospers or withers, or whether a particular person recuperates or dies. So, from this perspective, it's pointless to ask that ultimate power over all for preferential victory in war, for health or for rain.

Accordingly, the Greeks made no religious sacrifices to Fate, and the Hindus built no temples for Atman. And while the Aztec Empire obliged its subjects to build temples for the lead God Huitzilopochtli, they were frequently constructed alongside those of local gods, rather than in their stead. Many of the imperial elite honored those polytheistic gods and rituals.

Unlike monotheists, polytheistic empires generally didn't attempt to convert subjects of new territories they conquered or coveted. The Egyptians, the Aztecs and the Romans didn't send missionaries to foreign lands to spread the worship of Osiris, Huitzilopochtli or Jupiter—nor dispatch armies to enforce any such intent.

Early Romans readily adopted foreign deities such as the Asian goddess Cybele and the Egyptian goddess Isis to their pantheon. The monotheistic evangelization of Christianity presented a long-resisted special problem in this regard—one seen as disrespect for the empire's protector gods, disloyalty to the emperor and ultimately, a politically subversive danger.

Persecution by polytheistic Romans led to thousands of tragic Christian executions over the over the course of three decades. Still, these casualties were dwarfed over the course of the next 1,500 years by millions of Christians slaughtered by other Christians in theological disputes over slightly different interpretations of that same religion. Hundreds of thousands of Catholics and Protestants killed each other during just the sixteenth and seventeenth centuries alone.[lxxxiii]

Paradoxically, some ardent monotheistic religions, including Christianity, also sanctify de facto polytheistic beliefs and liturgies. Just as Jupiter defended Rome, and also as Huitzilopochtli protected the Aztec Empire, every Christian kingdom has enlisted divine help and protection from a particular patron saint assigned to their special causes.

England was protected by St. George, Scotland by St. Andrew, Hungary by St. Stephen and France by St. Martin. Cities and towns had their own saints also: Milan had St. Ambrose, while St. Mark watched over Venice.

Even various professions have had their own saints. St.

Larry Bell

Florian protected chimney cleaners, whereas St. Mathew lent a
hand to tax collectors in distress. If you suffered from
headaches you had to pray to St. Agathius, but if from
toothaches, then St. Apollonia was a much better choice.[lxxxiv]

An altogether different kind of religion began to spread
through Afro-Asia in the first millennium BC. These creeds,
including Jainism and Buddhism in India, Daoism and
Confucianism in China and Stoicism, and Cynicism and
Epicureanism in the Mediterranean basin put supreme faith in
natural laws. While some of these religions continued to
espouse the existence of gods, these deities were subject to
obeying the same laws that govern humans, animals and plants.

The central figure in Buddhism, the most influential of
these ancient natural-law religions, is not a god, but a human
named Siddhartha Gautama. Born around 500 BC as prince heir
to a small Himalayan kingdom, Gautama was affected by
suffering evident all around him. He observed that even those
who were rich and famous are rarely satisfied with their lives.

Prince Gautama traveled as a homeless vagabond
throughout northern India over six years, realizing in the
process that suffering is not caused by ill fortune, by social
injustice or by divine whims of gods. Rather, suffering and
discontentment arise from behavior patterns of one's own
mind.

Gautama established a set of ethical rules applying
meditation techniques that focus attention to high ethical
principles and away from cravings for power, sensual pleasure
or wealth. He taught that perfection of these spiritual
attainments leads to "nirvana" (literally meaning to extinguish
the fire).

According to Buddhist tradition, Gautama himself attained
nirvana and was fully liberated from suffering. Henceforth, he
became known as "Buddha," which means "The Enlightened

One."

The terms "Hindu" or "Hinduism" denote an unbounded set of theological beliefs with no compulsory dogmas that have persisted for nearly 4,000 years. Predicated on the idea that eternal wisdom of the ages can't be confined to a single sacred book, it embraces doctrines and practices ranging from pantheism to agnosticism, from faith in reincarnation to belief in the caste system.

Whereas other religions may look to find God in the heavens, the Hindu looks within themselves. There is no Hindu pope; no Hindu Vatican, no Hindu catechism; no prescribed divinities to adore or pray to; no religious Hindu rituals to honor or customs to practice; and no visible sign of Hindu identity to wear or display.

Given that Hindus make no common claim regarding what God looks like, individuals are free to imagine Him or Her as a woman with eight arms riding a tiger; as a pot-bellied man with an elephant's head; or as a muscular figure with a monkey's head and tail.

Hindu texts operate upon a foundation of skepticism regarding heavenly certitude dating back to a compilation of 3,500-year-old Rig Veda poems. The hymn "Nasadiya Sukta" addressing mysteries of creation concludes: "In the highest heaven, only He knows—or perhaps he does not know."

Whereas most faiths believe that the soul has a body, most Hindus believe that the soul has a temporary body which the Eternal Atman resides in, then discards. Some others dismiss any need to accept belief in the existence of any God, whatsoever.

Hinduism is perhaps the single major religion not claiming to be the only true religion. As declared by the noted 19th-century Swami Vivekananda, Hinduism teaches not only tolerance of other faiths, but acceptance of them as well.

Believers are encouraged to find their own answers to the true meaning of life through emphases on the mind, valuing reflection, intellectual inquiry and self-study.[lxxxv]

Establishment of Moral Principles and Laws

English novelist Aldous Huxley argued that there is a "perennial philosophy" or core of moral principles that exist in every time and place throughout history. Although variously interpreted through contemporary religious mandates and societal laws, some basic moral scaffolding appears to have been constructed upon common ancient foundations.

The "Code of Hammurabi" is one of the oldest recorded written laws containing a collection of 282 rules established by Babylonian king Hammurabi who reigned from 1792 to 1750 BC. Two earlier but less famous codes of conduct from the Middle East bear strong similarities: one created by the Sumerian ruler Ur-Nammu in the 21st century BC; the other a Sumerian Code of Lipit-Ishtar crafted two centuries before Hammurabi.

Hammurabi's Code, which was carved onto a massive black stone pillar, codified principles of justice and punishment which are both prescient and bizarre by today's standards. In the former category is a doctrine of "innocent until proven guilty," although with the extreme misfortune to any accuser who failed to make their case. Translated, it says, "If anyone brings an accusation of any crime before the elders, and does not prove what he has charged, he shall, if it be a capital offense charged, be put to death."

Another famous example is the ancient precept of retaliatory justice commonly associated "an eye for an eye," one often leading to grisly consequences. If a man broke the bone of one of his equals, his own bone would be broken in return. And sometimes there was far less equality in extracting reciprocal retribution for an offense. If a pair of scheming lovers conspired to murder their spouses, both were impaled, and if a son hit his father, the code demanded that the boy's hands be "hewn off."

The severity of a penalty depended upon the social class level of the lawbreaker and victim. For example, "If a man knocks out the teeth of his equal, his teeth shall be knocked out," whereas committing the same crime against a member of a lower class was punished only with a fine. If a man killed a pregnant "maid-servant," he was punished with a monetary fine, but if he killed a "free-born" pregnant woman, his own daughter would be killed. Men were allowed to have extramarital relationships with maid-servants and slaves, but philandering women were to be bound and tossed into the Euphrates along with their lovers.

Hammurabi's Code also mandated a pecking order of appropriate compensation standards for payment of services based on social class and occupation. A doctor's fee for curing a severe wound was set at 10 silver shekels for a gentleman, five shekels for a freedman and two shekels for a slave. Penalties for malpractice followed the same principle: a doctor who killed a rich patient would have his hands cut off, while only financial restitution was required for maltreatment of a slave.

The code also specified minimum wages. Field workers and herdsmen were guaranteed a wage of eight gur of corn per year, whereas ox drivers and sailors received six gur.

Hammurabi's Code remained influential in the region for centuries after his empire went into decline after his death in 1750 BC and crumbled completely when a Hittite army sacked Babylon in 1595. In the words of his monument, the purpose of his code was "to prevent the strong from oppressing the weak and to see that justice is done to widows and orphans."

According to one ancient Hindu creation myth, the gods fashioned the world out of the body of a primeval being (a cosmic man whose sacrifice by the gods created all life)—the Purusa. The sun was created from the Purusa's eye, the moon from the Purusa's brain, the Brahmins (priests) from its mouth, the Kshatriyas (warriors) from its arms, the Vaishyas (peasants and merchants) from its thighs and the Shudras (servants) from its legs.[lxxxvi]

In *Songs of Chu,* an anthology of Chinese poetry during the Han dynasty (340 BC-278BC), when the goddess Nüwa created humans from earth, she kneaded aristocrats from fine yellow soil, whereas commoners were formed from brown mud.[lxxxvii]

STRUGGLES FOR UNDERSTANDING

TRANSITIONS FROM AND between dogmatically metaphysical and empirically rational references of thought and behavior have been stutteringly erratic and turbulently brutal. Remarkably rapid intellectual advancements which began in Western Europe during the early centuries preceding the birth of Christ were fiercely resisted by religious hierarchies heading the early Church named in his honor.

Cruel penalties were inflicted upon those whose ideas were perceived to conflict with sanctioned articles of faith and dictates of all-powerful pontiffs. Bloody "holy" Crusade conflicts between the Church and non-compliant ideologies— as well as wars between various Christian sects—made free-thinking a perilous offense.

A Church-orchestrated and effectuated Inquisition exacted unthinkable terrors against those suspected of heresy. Included were young female children accused of witchcraft who suffered particularly gruesome fates.

Church corruption and populace control gradually gave way to a Renaissance emergence from a dark quagmire of anti-intellectualism that had smothered cultural and scientific advancement over nearly 200 years up until the beginning of the European High Middle Ages around 1000 AD.

Although empiricism continued to struggle against mysticism, there was and is no way to stop its determined strides of expansion and achievement. Followed and fueled by the Scientific Revolution and period of Enlightenment, civilization had chartered an open pathway to unlimited discovery, creativity and individual potential.

The endless expansion of human innovation, of course, long-predates medieval minds and European heritage. We've already seen that striking creative advancements were evident in tool-making during the Upper Paleolithic period which began about 40,000 years ago, the invention of agriculture in the Neolithic about 10,000 years ago and early Harappan urbanization which peaked in India about 2,000 years ago.

Chinese and Egyptian Pave Roads of Progress

Much progress which laid important paving stones for European scientific and cultural advancement can be credited to Chinese origins. Particularly consequential among these were contributions to horticulture, mechanical printing, means for precise ship navigation and gunpowder.

Chinese farming in southwest Asia's Fertile Crescent about 10,000 years ago led to the invention of grafting. The Chinese learned that trees grown from special cuttings produce reliably higher quality fruits such as apples, pears and cherries than those grown from seeds. This discovery contributed to an ability to feed their population, the world's largest at the time.[lxxxviii]

Also notable, it was the Chinese who invented movable

type, not Guttenberg who later developed the process in Europe where it had a greater revolutionary impact. Chinese applications of this printing method were made far more difficult due to an ideographic picture writing system involving myriad visual concepts rather than a comparatively very short and simple alphabetic iconography.

In combination with advanced printing technology, the Chinese also invented paper to go with it.

The Chinese discovered the circulation path of blood before English physician William Harvey, who is broadly credited, later did so in the 16th century. They also described the scientific first law of motion before English physicist and mathematician Sir Isaac Newton formulated quantitative proofs in the 17th century.[lxxxix]

The Chinese invention of magnetic compasses and mechanical clocks greatly influenced world exploration. Seafarers prior to these innovations navigated by dead reckoning and coastal guidance. For latitude, they relied upon the siting of Pole Star elevation above the horizon, yet because it vanished at the horizon, they then used the Sun's position in the southern hemisphere as a proxy. This presented a problem because the Sun's observed elevation shifted throughout the year, requiring calculation by charting tables.

Meanwhile, a more accurate location also required timing calculations of celestial movement in the longitudinal plane. Although early navigators correctly understood the time function, precise longitudinal measurements required non-existent accurate clocks.

Fifteenth-century Chinese explorers put these seafaring navigation advancements to ambitious use, which arguably eclipsed later European voyages of discovery. Between 1405 and 1433 AD, Admiral Zheng He of the Chinese Ming dynasty led seven huge armadas from China to the far reaches of the

Indian Ocean. The largest of these comprised almost 300 ships and carried close to 30,000 people.

Zheng He's sailors visited Indonesia, Sri Lanka, India, the Persian Gulf, the Red Sea and East Africa. His ships anchored in Jeddah, the main harbor of the Hejaz, and in Malindi, on the Kenyan coast. Christopher Columbus' fleet of 1492 consisting of only three small ships manned by 120 sailors was very tiny by comparison.

During the 1430s, a new ruling faction of Beijing overlords abruptly terminated Zeng He's operations. His great shipping fleet was dismantled, crucial technological and geographical knowledge was lost and no explorer of such stature and means ever set out again from a Chinese port.

The Chinese invention of gunpowder later had consequential world-wide military influences upon large and small feudal systems and empires alike. Together with horticultural advancements to support ever-growing populations; machine printing and paper, which extended public information access; the clock, which enabled precise navigation and changed the broad ordering of human activities, it impacted civilization in immeasurable ways.

Yet China's influence would have been much greater were it not for the country's voluntary withdrawal from the outside world and their leaders' suppression of free inquiry. Reliance upon mystical texts aggravated by ideographic written language and bureaucratic suppression of scientific inquiry between the 15th and 20th centuries contributed to China's demise as an invention capital.[xc]

China's example once again demonstrates the pervasively inhibitive influences of powerful hierarchical bureaucracies. Overseen by strong centralized authority, a huge civil service had been established to control the country's vast irrigation system.

Promoting severely stratified society, the system discouraged efforts to grow to high achievement levels. To this day, the East evidences far higher conformity to tradition and authority than does the West.[xci]

From Christ's time to the mid-14th century, the Chinese were prolific inventors, yet ultimately lagged the West in developing science. Here, abstract ancient Greek thinking corresponded more closely with creative incentives for theoretical scientists striving to advance knowledge beyond desolate frontiers of bureaucratic and dogmatic ignorance.

Critical thinking, which relegated ancient Greek gods to the background, also contrasted with conditions in Egypt. There, the most learned men became priests who suppressed seminal thinking which their orthodoxy considered to be heretical. Egyptian adherence to sacred beliefs, subservience to authority and commitment to major works of cooperation exemplified by pyramid construction prevailed, whereas ancient Greek efforts became more individualistic.

Nevertheless, the Egyptian industry benefited Greek intellectuals as well. By the middle of the 8th century BC, the ancient Greeks imported papyrus from Egypt. This transportable, writing substrate enabled the Greeks to more readily distribute commercial records and treatises on technical matters.

Surges of Creative Greek Genius

From about 2000 BC until the 2nd century AD, ancient Greece became a primary center of intellectual achievement. This began as Greek-speaking Bronze Age nomads from distinct but culturally-connected regions invaded lands of the Aegean in intermittent waves.

The new settlers gradually adopted a new sense of social consciousness which questioned many previously fixed ideas.

Although continuing to observe traditional religious customs and fertility rites, acceptance of human sacrifice by the ceremonial killing of prisoners declined.

Homer had completed his epic *Iliad* and the *Odyssey* poems by 800 BC, now regarded as central works of ancient Greek literature. The *Iliad* theme is set during the Trojan War, the ten-year-long siege of the city of Troy by a coalition of Greek kingdoms.

Whereas Homer probed the past for knowledge, later Greek philosophers looked to the future.

Nevertheless, primitive thought survived in the form of 4th and 5th century BC mythologies such as one attached to the legendary prophet, poet and musician Orpheus, who descended into the Underworld and returned—the source of Greek Orphic literary tragedy.[xcii]

By the mid-450s BC, Athens had grown to become a thriving commercial, maritime city where contact with wide-ranging cultures and world perspectives advanced conceptual thinking. Pericles, a great Athenian statesman, is credited as a powerful force in establishing the city as an important political and cultural empire. The Parthenon's construction in 447 BC is emblematic of Athens' renowned democratic stature.

Greek creative explosion represents one of humanities' giant leaps in human cognition. Gradually replacing anthropomorphic deities with natural phenomena, ancient Greek philosophers gradually intertwined naturalism with mythological thought. The imagined powers of gods waned as religious skepticism began to pervade Greek intelligentsia.

Rationalism rapidly came to the Greek mind by emphases on debate, politics and scientific schools in Ionia. Although the Greek majority continued to respect Olympian gods, Ionian scientists devoted their thoughts more an emphasis upon material substance which possessed neither divine order nor purpose.[xciii]

The epicenter for such discourse and debate was near the Ionian-dialect-speaking Greek city of Miletus, which prior to 500 BC, had become a major trade outlet for products from the interior of Anatolia in western Turkey and Sybaris, in southern Italy.

In addition to its commerce and colonization, Miletus was distinguished for its literary and scientific-philosophical figures. Revolutionary thinking seemingly occurred in Miletus with remarkable suddenness as Ionian philosophers replaced mythology with critical rationalism based upon speculations regarding natural phenomena as understood at that time.

Around 600 BC, Thales of Miletus made the seminal observation that the world possessed rationality and order. Initiating philosophy and science, Thales believed world mysteries could be discovered in thought. Whereas Thales erroneously proposed that all matter was composed of water as a single physical entity, he boldly and historically broke ranks with traditional beliefs in animism (a supernatural attribution of the soul to plants, animals and inanimate objects) and mysticism to explain natural phenomena.

Anaximander of Miletus, who wrote about biology, geography and cosmology, proposed a theory that the Universe was boundless and originated from the "aperion" (the "infinite," "unlimited" or "indefinite"), rather than from a particular element as water (as Thales held).

Perhaps the first true evolutionist, Anaximander believed that man could not have survived had he always been as he had become. He argued that the long period of care needed by human young precluded non-evolutionary views. Backed by observations of fossil remains, Anaximander contended man evolved from fish at sea.[xciv]

Ionian philosopher and scientist Pythagoras (circa 500-475

BC) made important contributions to mathematics and geometry which are believed to have later influenced Socrates, Plato, Aristotle and other early Greek thinkers. He is credited with many important discoveries which include the famous Pythagorean Theorem (a fundamental relation in Euclidean geometry), the five regular solid geometries, the Theory of Proportions, the Earth's spherical shape and the identity of the morning and evening stars as the planet Venus.

Of special note, although the Babylonians knew the Pythagorean Theorem at least a thousand years before Pythagoras, it was he who conceived the mathematical proof—that for a right-angled triangle the square of the hypotenuse equals the sum of the squares of the other two sides.

Pythagoras was also devoted to religious mysticism and a believer in the "transmigration of souls," wherein every soul is immortal and, upon death, enters into a new body.

Greek philosopher Anaxagoras of Clazomenae (a major city in Ionian Asia Minor) was first to give a correct explanation of eclipses and was both famous and notorious for theories, including the claims that the Sun is an incandescent stone (not a god), and that the Moon consists of Earthly materials. Yet while disparaging the idea of mythology and anthropomorphic gods, Anaxagoras believed that a transcendent mind set the world in motion, giving it form and order.

Philosophers Leucippus and Democritus developed the idea that the world consisted exclusively of invisible, minute, uncaused and immutable material called atoms. Moving in a boundless void, atoms occasionally collided and combined to produce the visible world. They theorized that this ceaseless movement arose mechanically, not by divine guidance.

Heraclitus and Parmenides posed contrasting thoughts in pre-Socratic Greece where synthesis of opposing views was recognized often to produce synergistic order which was often greater in merit than either extreme.

The philosopher Heraclitus, who lived in Ephesus, a city on the Ionian coast not far from Miletus, offered new meaning to "Logos." Opposed to the original meaning—word, speech or thought, he considered Logos divine intelligence or "the rational principle governing the Universe." Heraclitus postulated that while things are in constant flux, they are also related and ordered by universal Logos.

Heraclitus disbelieved permanent reality, that all things are eternally transient and motile. He reasoned that most humans exist in false dreams and conflict, but their understanding of Logos enables harmonization with deeper realities.

Parmenides from Elea, a Greek city in southern Italy, believed all things consisted of immutable, elementary particles, motion seen as sense illusion.[xcv]

Hippodamus of Miletus is credited as one of the first political theorists who believed that the emergence of the "polis," the Greek city-state and philosophy occurred in concert.

Ancient Greeks considered public life critical to human activity, whereby mythology became dissipated in concert with their belief in the equivalency between principles of reason and politic. Emerging separately from religion, political thought promoted secularism, and reason was viewed as the inherent responsibility and right of free men which fundamentally shaped pre-Socratic thinking in all areas of life.[xcvi]

A Sophist movement led by Protagoras in the mid-400s BC believed in independent thinking, that "Man is the measure of all things" and that each person's opinion, therefore, possessed a level of truth. This came to take on a Machiavellianism emphasis, which as a consequence, defined justice being relinquished to the advantage of the stronger.[xcvii]

Although Protagoras has come to be regarded as a true and sincere pragmatist, later Sophists lost that intellectual honesty. They instead promoted the politically weaponized use of rhetoric—eristic as oppose to dialectic. Eristic is deception by seemingly validating falsehoods.

Sophist students became skilled in the art of eristic, or how to devise plausible arguments supporting untrue claims...today called "spin."

The Sophists viewed speculative philosophical theories regarding cosmologies and speculative science implausible and useless. Critics including Aristophanes, Plato and Aristotle later ridiculed this philosophy as dishonest, unethical and shallow, leading to sophistry's derogatory meaning.

The Sophists' relativistic humanism led to broad Athenian skepticism toward all ethical values. Many within the populace found this view lacking and disruptive to social and moral good.

Whereas pre-Socratic natural philosophers had spurned religion and sought material explanations for phenomena, later philosophers—most notably Socrates, Plato and Aristotle—combined mysticism with naturalism. Accordingly, a dichotomy between mysticism and rationalization characterized Greek thought from the time of Socrates (mid-300s BC) and onward.

During these times of tension, and perhaps to a large degree in reaction to failures of Sophist philosophy, Socrates sought a system of inquiry based upon earnest dialog which was both rational and emotional; a pursuit of knowledge where "truth" was attainable, yet to be achieved. He also believed that seeking knowledge followed a sacred pathway to God.[xcviii]

Socrates' last dialogue, *Phaedo*, attempted to prove the human soul is immortal. Philosophically most important in this was his description of hypothesis and deduction separate from experiment and observational proofs.[xcix]

In 399 BC, Socrates, then 70 years old, stood before a jury of 500 of his fellow Athenians accused of "refusing to recognize the gods recognized by the state" and of "corrupting the youth." If found guilty, his penalty could be death. His anti-democratic views had turned powerful authorities and politicians against him, particularly after two of his students, Alcibiades and Critias, had twice briefly overthrown the city government.

After hearing arguments of both Socrates and his accusers, the jury found him guilty by a vote of 280 to 220. Given an opportunity to suggest his own punishment, Socrates might have avoided death by recommending exile. Instead, he initially offered a sarcastic proposal that he be rewarded for his actions. When pressed for a realistic punishment, he next offered to pay a modest fine.

The jury selected a death penalty, and Athenian law prescribed that this be carried out by drinking a cup of poison hemlock. Thus, Socrates became his own executioner, reportedly doing so with equanimity.

Socrates' student Plato later (428-347 BC) founded the Academy in Athens, the first institution of higher learning in the Western world. Plato (meaning "broad" referring to his broad shoulders and forehead) was actually a pseudonym for his real name, Aristocles.

Upon Socrates' execution, Plato lost trust in Athenian democracy—law without standards of justice. Holding to the concept of a supreme ruler, Plato believed that it is essential to establish a standard foundation for philosophies and political systems—one of "absolutism."

Fundamentally a rationalist, Plato's dialogues reflect a dualism between the spiritual and rational. In his "Republic," he emphasized dialectic and rigorous self-critical logic. In his "Symposium," Plato championed the supernatural.

Plato proposed geometrical atomism: reality arises from

mixtures of space and form. In his view, four elements comprised the physical and biological world: the basic particles of fire, Earth, air and water.

Prominent intellectual Greek thinking at the time viewed the heavens as timeless and unchanging with planets directed by spiritual powers traveling in perfectly circular orbits.

A firm belief in divine, circular order caused Plato to consider it blasphemous to refer to planetary irregularities and their multiple wanderings. He believed critical mathematical reasoning alone would reveal perfect circular orbits of planets, and "the Universe as the living manifestation of divine Reason."

Starkly contradicting his teacher Plato, Aristotle (384-322 BC) believed that the validity of all philosophical theories must be supported by evidence. Aristotle understood—as others before him had not—that deduction alone is inadequate for achieving reliable knowledge.

Although Aristotle recognized that empirical knowledge is also fallible, he believed that ultimate divine truth could be discovered through active intellect which combines empiricism with rationalism.

Aristotle's science and philosophical thinking applied a disciplined and systematic concept of logic aimed at providing a universal process of reasoning that would allow humans to learn every conceivable thing about reality. He characterized deduction as arising from a reasonable argument in which:

> [W]hen certain things are laid down, something else follows out of necessity in virtue of their being so.

This precept became the basis for what philosophers now refer to as a "syllogism," where a logical conclusion is inferred from two or more other premises of very particular forms. This view was at odds with Plato's teachings, which noted deduction simply follows from premises, which lead to seemingly logical conclusions.

At the age of 17, Aristotle had left his birthplace in the small town of Stagira on the northern coast of Greece to enroll in Plato's Athens Academy. Following Plato's death in 348 BC, his friend Hermias, King of Atarneus in Mysia, invited Aristotle to his court where he met and married the king's niece Pythias.

Seven years later, King Philip of Macedonia handsomely compensated Aristotle to tutor his son, the then 13-year-old Alexander, who became a close friend. Aristotle subsequently became a key and an extremely harsh military counselor during Alexander's Eastern conquest of Persia. He counseled Alexander to be:

A leader to the Greeks and a despot to the barbarians, to look after the former as after friends and relatives, and to deal with the latter as with beasts or plants.

Aristotle's subsequent appointment to head the Royal Academy of Macedonia also involved giving lessons to two other future kings, Ptolemy and Cassander.

Aristotle returned to Athens in 335 BC after Alexander, who succeeded his father, King Philip, conquered the city. Since Plato's Academy was then being headed by Xenocrates, with Alexander's blessings, Aristotle created his own school called the "Lyceum." There, he was believed to be the first teacher to organize his lectures into courses and to assign them a place in a syllabus.

Building upon principles worked out by Socrates and Plato, Aristotle developed incipient scientific thought that became a foundation for modern inductive and deductive logic and natural science. Although much of his science was later disproved, he established a basis from which mistakes could be recognized and more reliable knowledge derived.

Aristotle divided the sciences into three general categories: productive, practical and theoretical. The productive sciences not only included engineering and architecture, which yield tangible products such as bridges and houses, but also disciplines such as strategy and rhetoric, where a product is something less concrete, such as victory on the battlefield or in the courts.

Practical sciences, most notably ethics and politics, were those associated with guiding behaviors. Theoretical science involved those that have no product and no practical goal, but in which information and understanding are sought for their own sake. He divided these into three groups: physics (the study of nature), mathematics and theology.

Aristotle's political relationships during those turbulent times brought hard consequences. His relationship with Alexander cooled as the king became more and more megalomaniac, proclaimed himself divine and demanded that Greeks prostrate themselves before him in adoration.

Aristotle was later charged with impropriety for his previous relationship with Alexander's government following the sudden death of Alexander the Great and overthrow of his reign of power in 323 BC. He then fled to Chalcis in 321 BC to escape prosecution and execution and died a year later by drinking a poisonous hemlock beverage. He'd been charged with the same offenses that ended the life of Socrates.

In death, Aristotle left behind an estimated one million words of his surviving works recorded on papyrus scrolls. These scrolls are estimated to represent around one-fifth of his total output. Although his writings generally fell out of use soon afterwards, their retrieval more than seven centuries later significantly influenced Western thought on humanities and social sciences.

Whereas Plato and Aristotle are both ranked by many historians as being the greatest philosophers who have ever lived, it is Aristotle who might better be credited for his contributions to intellectual empiricism, as his work inspired other great thinkers, including Leonardo da Vinci during the Renaissance.

Altogether, the ancient Greeks rapidly achieved a remarkable quantity and quality of intellectual advancements which did not reappear again until the Renaissance and the Scientific Revolution some 1,400 years later. As Philosopher Bertrand Russell noted:

> *Within the short space of two centuries, the Greeks poured forth in art, literature, science and philosophy an astonishing stream of masterpieces and set the general standards of Western civilization.*"[c]

Briefly summarized, a short sampling of these contributions includes:

- Pythagoras, (582-507 BC): recognized that the Earth moves.
- Eudoxus, (408-355 BC): contributed to mathematical astronomy; calculated approximate planetary positions.
- Heraclides, (390-322 BC): proposed that the Earth's rotation causes diurnal movement and that Mercury and Venus always appear close to the Sun because they revolve about the Sun.

- Aristarchus of Samos, (310-230 BC): hypothesized that all planets revolve around a stationary Sun.

- Euclid, (circa 300 BC): made original contributions in natural science; his treatment of elementary plane geometry continues in use today.

- Archimedes, (287-212 BC): a mathematical physicist who contributed many aspects of physics; best known for his principle that a body immersed in a fluid is buoyed up by a force equal to the weight of the displaced fluid.

- Strabo, (63 BC-21 AD): an early geographer who wrote extensively about people and places.

- Galen, (130-200 AD): a physician who showed that arteries carry blood, not air; also advanced knowledge of the brain, nerves, spinal cord and pulse.

- Copernicus, (1473-1543): proved Aristarchus was correct when he developed his heliocentric theory.

Greek-Roman Era of Turmoil and Recovery

The age of ancient Greek inspiration declined beginning with 2nd-century conquest and domination by the 2nd century BC conquest by the Roman Empire. Although this new Roman era began with Grecian defeat in the Battle of Corinth in 146 BC, the Roman Republic had been steadily gaining control over the mainland following its successes in a series of conflicts known as the Macedonian Wars.

Definitive Roman occupation of the Greek world was established following the Battle of Actium (31 BC) in which Augustus defeated Egypt's queen, Cleopatra VII, and then soon afterward conquered Alexandria, the last great city of Hellenistic Greece.

The Roman era of Greek history continued with Emperor

Constantine the Great's adoption of Byzantium as "Nava Roma," the capital city of the Roman Empire. In 330 AD, the city was renamed Constantinople, with a general Greek-speaking polity.

Roman domination of lands of conquest adopted a culture of total subordination to Roman Christian orthodoxy with fierce suppression of heretical noncompliance with authoritative doctrine posed by scientific inquiries.

While the famous story of Cleopatra and her lover Marcus Antonius (Mark Antony) immortalized by Shakespeare and other writers may not have monumental historical consequences, it does reveal enormous political turmoil at the time. Born in 69 BC in the city of Alexandria, Egypt, Cleopatra had also been an ally and lover of Julius Caesar until the time of his assassination in Rome in 44 BC.

Caesar's death split Rome into various factions competing for control, the most important of these being armies of Mark Antony and Caesar's former supporter, close friend and later adopted son, Octavian.

In 41 BC, Mark Antony summoned Cleopatra to Tarsus (in modern southern Turkey). She majestically complied by sailing up the Cydnus River in an opulently decorated barge dressed in robes associated with the Greek goddess Aphrodite. Antony and Cleopatra immediately formed a legendary romantic and political liaison that later ended very badly for both of them.

The union between Cleopatra and Antony provided mutually prized strategic advantages. For Cleopatra it afforded an opportunity to achieve greater power in both Egypt and Rome; for Antony, it represented financial and military support from one of Rome's largest and wealthiest states in his campaign against the mighty Parthians (Parthia was a region in modern northeastern Iran).

The alliance between Cleopatra and Antony was not to be trifled with. She allegedly enlisted his support in successfully executing her half-sister Arsinoe, who had been living in the protection of the Temple of Artemis at Ephesus, to prevent any attempts against Cleopatra's throne. Antony and Cleopatra very publicly traveled to Alexandria together in 40 BC.[ci]

Despite an extravagant failure of his Parthian campaign, Antony and Cleopatra celebrated a mock "Roman Triumph" in the streets of Alexandria and issued a proclamation (known today as the "Donations of Alexandria") which caused outrage among Octavian's minions in Rome. The declaration (in 34 BC) purported to distribute lands held by Rome and Parthia amongst Cleopatra and her three children fathered by Antony.

To avoid civil war, Antony was not mentioned in the proclamation. Nevertheless, to complicate matters more dangerously, Antony then left Alexandria, traveled to Italy, and, presumably to cool down a rapidly heating conflict with Octavian, married his nemeses' sister, Octavia. This occurred concurrently at the time of his publicly flaunted celebrity relationship with Cleopatra.

Ongoing hostility erupted into civil war in 31 BC as the Roman Senate under Octavian's direction declared war on Cleopatra and proclaimed Antony a traitor. Later that year, Antony's naval forces were defeated at the Battle of Actium. Octavian then invaded Egypt in 30 BC, laying siege to Alexandria. Hopelessly outnumbered, Antony surrendered, and following Roman tradition, committed suicide by falling on his sword.

Following Antony's death, Cleopatra was taken to Octavian, informed that she would be brought to Rome, paraded through the streets, and following public humiliation, likely executed. According to the ancient historian Plutarch, Cleopatra instead arranged to have a poisonous asp (an

Egyptian cobra) brought to her concealed in a basket of figs and intentionally died from its bite along with two female servants. (Other historians, including Joyce Tyldesley, believe that Cleopatra used either a poisonous ointment or vial to commit suicide.) [cii]

In any case, the deaths of Antony and Cleopatra (the last monarch of the Ptolemaic Empire) left Octavian the undisputed master of the Roman world.

In 330 AD, the Roman emperor Constantine dedicated a "New Rome" (which he later named Constantinople) at the site of the ancient Greek colony of Byzantium located on the European side of the Bosporus linking the Black Sea to the Mediterranean—an ideal trade point between Europe and Asia. Five years later, at the Council of Nicaea, Constantine established Christianity as Rome's official religion.

Being monotheistic, Christianity undermined religious traditions of the Roman state which regarded the emperor as a god. This, in turn, tended to weaken the emperor's public authority and credibility.

In 364, following Constantine's death in 337, Emperor Valentinian I divided the Roman empire into western and eastern sections, putting himself in power in the West, and his brother, Valens, in the East. The western part of the empire spoke Latin and was Roman Catholic, while the eastern part spoke Greek and worshiped under the Eastern Orthodox branch of the Christian church.

It is important to note here that references to the subsequent "fall of Rome" relate only to the western part of the empire. The eastern half of the empire, the "Byzantine Empire," continued to exist for centuries.

Issues other than religious disputes contributed to the fall of the western part of the original Roman Empire. A decrease in agricultural production led to higher food prices, which ultimately led to a large trade deficit with the eastern half. Whereas the west continued to desire eastern luxury goods, they had nothing to exchange.

To make up for a lack of money, the western government began to produce coins with less silver content, which led to inflation. Adding to this problem, piracy and attacks from Germanic tribes disrupted the flow of trade with particular pain to the west.

Wave after wave of Germanic barbarian tribes, including Visigoths, Vandals, Angles, Saxons, Franks, Ostrogoths and Lombards, swept through and ravaged the Roman Empire. Romulus Augustulus, the last of the western Roman emperors, was overthrown by the Germanic leader Odoacer in 460 AD. Odoacer became the first Barbarian to rule Rome.

The order that the Roman Empire had brought to Western Europe over 1,000 years had ended.

Meanwhile, the Eastern Roman Empire (variously known as the Byzantine Empire, or Byzantium) tended to flourish. Although ruled by Roman law and political institutions with its official language as Latin, Greek was also widely spoken, and Greek history, literature and culture were included in education systems.

In 451, the Council of Chalcedon officially established a division of the Christian world into five patriarchates, each ruled by a patriarch: Rome (where the patriarch would later be called pope), Alexandria, Antioch, Jerusalem and Constantinople (where the patriarch emperor would head both church and state).

Even after the Islamic Empire later absorbed Alexandria, Antioch and Jerusalem in the seventh century, the Byzantine emperor would remain leader of most eastern Christians.

Justinian I took power over the Byzantine Empire in 527 and controlled most of the land surrounding the Mediterranean

Sea, including part of the former Western Empire (such as North Africa) captured by his armies. Justinian established a Byzantine legal code that endured for centuries, and built great monuments including the spectacular domed Hagia Sophia (Church of Holy Wisdom).

By the time of Justinian's death in 565, the Byzantine Empire reigned as the largest and most powerful state in Europe. At the same time, however, debts incurred in wars had left the empire in dire financial straits. As a result, Justinian's successors were forced to levy heavy taxes upon Byzantine citizens to keep the empire afloat.

Making matters worse, Justinian's vain struggles to maintain control of his territories had left the imperial army stretched very thin. Seventh and eighth-century attacks from the Persian Empire and from Slavs combined with internal political instability and economic regression threatened the empire's survival.

A new and more ominous threat arose in the form of Islam, a religion founded by the prophet Muhammad in Mecca in 622. Muslim armies soon began to attack the Byzantine Empire by storming into Syria.

Eighth and ninth century Byzantine emperors (beginning with Leo III in 730) began to deny homage of holiness to all icons of non-Christian worship. Referred to as "Iconoclasm"— literally "the smashing of images"—the movement waxed and waned under various leaders until 843 when a Church council under Emperor Michael III ruled in favor of religious displays.

The late 10th and 11th centuries enjoyed a golden age of Macedonian dynasty art and culture initiated by Michael III's successor, Emperor Basil. Although his Byzantium stretched over less territory, it controlled more trade, wealth and prestige than under Justinian.

Macedonian rulers began restoring churches, palaces and

other cultural institutions. They also promoted the study of ancient Greek history and literature, establishing Greek as the official state language.

Monks administered many institutions (orphanages, schools and hospitals) in everyday life, and missionaries won many Christian converts among Slavic peoples of the central and eastern Balkans (including Bulgaria and Serbia) and Russia.

Killing for Love of God and Neighbor

The golden age ended when the late 11th century witnessed the beginning of a series of holy wars waged by European Christians against Near Eastern Muslims between 1095 and 1291. Termed the "Crusades," the conflicts began when Constantinople Emperor Alexius I turned to the Western Empire for help to defend against attacks from Central Asian Seljuk Turks.

The Crusades also arose partly as a consequence of overpopulated warrior class in Europe, causing local warfare. Population growth led to unrest and soil infertility. This unrest became aggravated by the requirement that younger sons were rejected from land that only eldest sons inherited.[ciii]

The Church promoted taking the cross as a demonstration of "love of God" and "love of neighbor." Likewise, knights were taught that to be a good Christian knight required undertaking acts of such love and charity in God's name.

Commenting on Muslim victories in the holy land, French monk Bernard of Clairvaux wrote:

> *If we harden our hearts and pay little attention...where is our love of God, where is our love for our neighbor?*

To maintain a hold over warrior aristocracy, a papacy fearful of losing power directed warfare. As a huge incentive, the Church taught at that time that an individual's sins could be remedied, at least in theory, by acts of penance that demonstrated remorse and a desire for forgiveness. As communications through Central Europe improved, and Italian trade in the Mediterranean increased, more Western European people than ever before could journey or make a pilgrimage to the Holy Land and seek penance for past sins.

Pope Urban II responded to Emperor Alexius' call for armed assistance from Turk invasions by declaring a "holy war" which commenced the First Crusade (1096-1102). As a result, the Europeans captured Jerusalem in 1099. The Muslims, however, quickly unified against the Christian invading and occupying force, and the two groups battled in subsequent wars for control of the Holy Land. By 1291, the Muslims firmly controlled Jerusalem and the coastal areas, which remained in Islamic hands until the twentieth century.[civ]

Christians were promised that the Crusades offered a path to salvation for those who participated. As the 12th-century French monk Guilbert of Nogent instructed:

> *God has instituted in our time holy wars, so that the order of knights and the crowd running in its wake...might find a new way of gaining salvation. And so they are not forced to abandon secular affairs completely by choosing the monastic life or any religious profession, as used to be the custom, but can attain in some measure God's grace while pursuing their own careers, with the liberty and in the dress to which they are accustomed.*

Those who "took up the Cross" were to become recipients of

both spiritual and Earthly rewards, in addition to gaining indulgences—forgiveness of sins—they were also to receive shares of plunder from conquests, forgiveness of debts, freedom from taxes, fame and privileged access to political influence.

As a result, millions of people, Christians and non-Christians, soldiers and noncombatants lost their lives. In addition, the debt incurred along with other economic costs associated with multiple Middle East excursions impacted all levels of society—from individual families—to the many villages and cities that were destroyed in the Crusaders' wake.

Altogether, Christians made eight unsuccessful attempts, including an unsanctioned Children's' Crusade, to recapture the Holy Sepulcher, the Holy Grail and the Holy Land from the Muslims.

In his book *The History of the Rise and Fall of the Roman Empire,* eighteenth-century writer Edward Gibbons refers to these Crusades as events in which "the lives and labours of millions, which were buried in the East, would have been more profitably employed in the improvement of their native country." [cv]

Approximately beginning with the reign of Emperor Michael VIII in 1261, the economy of the once-mighty Byzantine state was crippled, never to regain its former stature.

Emperor John V Palaiologos, the son of Emperor Andronikos III, oversaw the gradual dissolution of imperial power amid numerous civil wars and witnessed continuing ascendancy of the Ottoman Turks.

In 1367, John V appealed to Pope Urban V in the West for help to stave off the Turkish threat and to end a schism between the Byzantine and Latin churches by voluntarily submitting the patriarchate to the supremacy of Rome. In 1369, he traveled through Naples to Rome and formally converted to Catholicism in St. Peter's Basilica where he recognized the pope as the supreme head of the Church. This concession was not enough, however, to end the split.

Impoverished by wars, John was detained as a debtor in Venice on his return trip from Rome and was later captured on his way back through Bulgarian territories. In 1371, he submitted to the authority of Ottoman sultan Murad I.[cvi]

As a Turkish vassal state, Byzantium paid tribute to the sultan and provided him with military support. Although the Byzantine Empire gained sporadic relief from Ottoman oppression under John's successors, the rise of Murad II as sultan in 1421 marked the beginning of the ending of a final respite.

Murad revoked all Byzantine privileges and laid siege to Constantinople. Murad's successor, Mehmed II, launched the final turnover of Constantinople to Turkish dominance when he triumphantly entered the Hagia Sophia and converted it to the city's leading mosque.

The fall of Constantinople in 1453 marked the end of a once-glorious Byzantine Empire, ushering in the long reign of the Ottoman Empire.

The centuries leading up to the final Ottoman conquest left a rich Byzantine legacy of philosophy, literature, art, science and invention which had flourished even as the empire itself had faltered.

Long after its end, Byzantine culture and civilization continued to exercise an influence upon countries that practiced its Orthodox religion, including Russia, Romania, Bulgaria, Serbia, Greece and others.

This broad and deep heritage came to exert particularly great influence on Western intellectual thought and traditions as Italian Renaissance scholars sought help from Byzantine scholars who had fled Constantinople for Italy in translating Greek and Christian writings.

Clouded Dawn of Renaissance

A Renaissance period dawn, which began in 14th century Italy, witnessed the emergence of numerous freethinkers, humanists and scientists who openly declared opinions opposed to religious ecclesiastical sanction. In parallel, that propensity for opposition to liturgical doctrine ended very badly for some of them as zealous tribunals exacted harsh penalties for suspected heresy.

Originally seen a pathway to the transcendental world of God, free-thinking scholastic rationalism, which set the stage for this Renaissance, had begun to upset Church dogma centuries earlier.

Transitional to the Renaissance, the Roman Church had become rife with corruption. By the 13th century, simony, the selling benefices to the highest bidder, was a common practice which heaped huge revenues into Church hierarchy coffers.

A political vacuum left by Imperial Rome transferred great power to popes who enacted doctrines which were anything but Christ-like. A long era of deep Church corruption and cruel heresy purges institutionalized torture, mutilation and incineration which continued to into the 17th century High Renaissance:

> *Pope Gregory VII (1020-1085) declared that "foul plague of carnal contagion" was responsible for clergy venality and authorized the laity to withdraw support from priests who did not renounce their wives and children. Abandoned by their husbands, church wives and children became destitute and starving. Under Gregory, simony became even more pronounced. Gregory also curtailed*

intellectual freedom, insisting that education was solely the Church's responsibility.

Pope Innocent III (1198-1216) exacted torture for women and freethinkers in Southern France.

In 1233, Pope Gregory IX formed a formal Catholic Church inquisition tribunal which sanctioned torture to obtain heretical confessions. In 1229, the Inquisition set up courts in Toulouse, rounded up many women accused of heresy, racked them until confession and burned them publicly.

In 1478, Spain's King Ferdinand and Queen Isabella appointed a cruel Dominican, Friar Torquemada, as Chief Inquisitor. He sanctioned torture of girls as young as 13, then sentenced them to die by burning.

In 1542, Paul III assigned the Inquisition to the Holy Office. Pope Paul IV who followed (1555-1559) terrified Italians by establishing severe religious persecution, removed nude paintings and statues, painted over Michelangelo's indiscretions in the Sistine Chapel, and burned thousands of books.

Galileo was famously placed under house arrest and brought before the Inquisition in 1630 for arguing against the geocentric model concept of Ptolemy backed by the Roman Catholic Church that the Sun orbited around the Earth (and not the other way around). Galileo's works advocating Copernicanism were then banned, and a papal order prohibited him from teaching, defending or even discussing his discovery.

In Germany, Kepler's works were also banned by the papal order.

Whereas astronomy gained momentum during the 12th century through assimilation of Greek and Islamic knowledge, Church-sanctioned scholars continued to believe that the heavens are spheres with circular motion, with Earth as the central body. This Earth-centrality could not be adjusted to correlate with the cycles, epicycles and eccentric of Ptolemaic astronomy.

Scientific inquiry was regarded as a curse by those who feared challenges it posed to Church doctrinaire authority. For example:

> *Medical practices based upon meticulous observation demanded by ancient physician Greek Hippocrates had become replaced with trust in miraculous healing which employed prayer, chants, potions, horoscopes and amulets. Disease was explained as the divine punishment for sin.*[cvii]

The Paris medical faculty proclaimed that a disastrous siege of Black Death (bubonic plague) between 1347 and 1351 was caused by "corruption of the air" following from the conjunction of Jupiter, Saturn and Mars.

The anatomical study of cadavers (other than those of criminals following execution) was forbidden by the Church until the 14th century.[cviii]

Church glorification by building beautiful, artistic structures such as St. Peter's Basilica was financed by sales of spiritual indulgences—payments for the remission of punishment for sins which also funded the Crusades. This corrupt practice led the German Augustinian monk Martin Luther (1483-1546) to rebel and break away from the Catholic hierarchy.

Luther's Protestant Reformation movement initiated the return to a fundamentalist Old Testament biblical emphasis which divided European Christianity into Catholic and Protestant followers.

Beginning in the late 1400s, violent persecution erupted between Catholics and Protestants, and also among Protestant sects. Religious wars waged in the name of God ravaged on in Europe for 150 years.

Thirteenth century Dominican Albertus Magnus (Albert the Great) was among the first medieval thinkers to distinguish between theological knowledge and science. Albert's broad interests included prolific writings about botany, astronomy, chemistry, physics, biology, logic, metaphysics, meteorology and zoology. His premise that faith and reason are not incompatible sources of knowledge inspired the major work of his most famous pupil, his Dominican colleague and friend, Thomas Aquinas.[cix]

Thomas Aquinas (1225-1274) fully amalgamated Aristotelian physics with Church doctrine. Like Albertus Magnus, he believed that combined thought forms faith and rationalism and benefited the Christian cause by arousing deeper understandings to support beliefs.

Also like Magnus, Aquinas was an intellectual who merged Greek with Christian ideals into a single holistic doctrine called "Summa Theologica." Aquinas agreed with Socrates—that evil stems from ignorance, less from malice—and his version of Aristotle's work came to be established as a dominant philosophy.

The Church canonized Aquinas as a scholar-saint in the

mid-13th century.[cx]

Whereas Franciscan scholars had argued for a sharp distinction between natural philosophy and theology, in their success, rational inquiry severed from theology resulted in a great boon to science. Relieved of religious restraints, free inquiry allowed those of faith to undertake unfettered scientific research.

Franciscan scholar Roger Bacon (1220-1292), who is celebrated as the first true experimental scientist in Britain, wrote in his "Opus maius" (Experimental Science):

> *Having laid down fundamental principles of the wisdom of the Latins so far as they are found in language, mathematics, and optics, I now wish to unfold the principles of experimental science, since without experiment nothing can be sufficiently known.*

Imprisoned for 15 years, Bacon's ideas were viewed unfavorably by the Church—empirical emphasis over metaphysical speculation was viewed his greatest sin.[cxi]

An acceptance of "scholasticism" gradually arose with Roman Catholic Church doctrine which held that God and religion could be justified by rationalism, a contention that natural knowledge offers a legitimate path to religious contemplation and mystical ecstasy.

British philosopher and priest William of Ockham emphasized the attainment of knowledge through logical thinking. Ockham established the philosophical principle that "entities are not to be multiplied beyond necessity." Known and still popular today as "Ockham's razor," it teaches that until proven otherwise, the preferred choice of answer to an uncertainty is the simplest explanation that best fits the facts.[cxii]

Intellectual growth inspired by scholasticism, along with reinstatement of Greek knowledge, established a springboard for great scientific and creative achievements associated with Renaissance genius. As explained by Richard Tarnas:

> *The medieval gestation of European culture had approached a critical threshold, beyond which it would no longer be containable by old structures.*

Nevertheless, the community of scholars remained to be polarized between opposing penchants of faith-revelation and evidence-logic.[cxiii]

The free-thinking Renaissance which emerged in the 14th peaked in the 15th and 16th centuries and ended in the 17th century as zealous religious police reinstated witch hunts for unbelievers. Luther's religious conservatism Reformation movement again separated theology from science whereby faith and reason clashed anew.

Free thinking attached to 16th and 17th century scientific and technological advancements provoked religious backlashes. Nicolaus Copernicus's heliocentric astronomical model published in 1543 which placed Earth and other planets along circular paths around a motionless Sun disrupted the concept of a human-centric, therefore church-centric, Universe.

Johannes Gutenberg's 15th-century invention of movable type printing along with innovations in rapidly casting type in hand molds dispersed ideas regarded to be intimidating to religious authorities beyond their control. Such backlash provoked renewed suppression of free-thinkers, which in turn stimulated even greater intellectual curiosity and determination.

Italian philosopher and mystic Giordano Bruno (1548-1600) was burned at the stake by the Inquisition, not as a martyr for science, but as a defender of his own unique philosophy of mysticism. Accused of diabolic practices, it appears that his unforgivable sin was to deny Christ's divinity.

Above all, the Renaissance is remembered as a period of creative genius, an era characterized by humanist literary and artistic expression immortalized by Petrarch, Boccaccio, Bruni, Alberti, More, Machiavelli and Montaigne in the early period—later by masterworks of Shakespeare, Cervantes, Michelangelo, Raphael and Leonardo da Vinci.

Among all of these creative giants, Leonardo da Vinci epitomizes the Renaissance ideal of a "Universal genius." Still widely recognized as one of the greatest artists of all time, he also engaged and excelled in an amazing variety of engineering and scientific endeavors ranging from machines for military defense and warfare, architecture and cartographical mapping, innovative concepts for flight and detailed studies of human anatomy and botany.

Leonardo's remarkable diversity recognized none of today's prevalent boundaries and mutually-exclusive polarities between science and art. His successes in both types of endeavors drew upon broad curiosity about the natural world which was combined with highly developed observational skills and dedicated attention to technical details.

Unlike Michelangelo, Raphael and most other religious artists of his time, Leonardo saw the world as logical rather than mysterious. He was also a fundamentally different kind of scientist than Galileo, Newton and others who followed him. Although Leonardo's observational approach to science attempted to understand a phenomenon by describing and depicting it in utmost detail, it did not emphasize experiments or theoretical explanations.

Consequently, while many of Leonardo's concepts either weren't technically feasible due to lack of scientific information, others were so ahead of his time that they required modern metallurgy and engineering advancements which only occurred centuries later. Examples of both include a fundamentally infeasible flapping "ornithopter" flying machine, and another with a helical rotor. His concept for armored double-hull ships used for military and commercial applications was successfully adopted long after his death.

At the same time, some of da Vinci's numerous inventions were not only implemented but were highly beneficial in his day. His designs for an automated bobbin winder, and a machine for testing the tensile strength of wire, proved revolutionary even then.[cxiv]

Leonardo was born out of wedlock (1452) in the Vinci region of the Medici-ruled Republic of Florence, home to many great artists and philosophers. Luminary figures included painters Piero della Francesca and Filippo Lippi, sculptor Luca della Robbia and writer and architect Leon Battista Alberti.

Such influential leaders were followed by Leonardo's teacher, the renowned artist Andrea del Verrocchio, Antonio del Pollaiuolo and painter-sculptor Mino da Fiesole. Leonardo was a contemporary of Botticelli, Domenico Ghirlandaio and Perugino...and most notably, Michelangelo and Raphael.

While all three of these High Renaissance masters were contemporaries of one another, they were not actually of the same generation. Leonardo was twenty-three years old when Michelangelo was born and was thirty-one when Raphael was born. Raphael died at age 37 in 1520...the year after Leonardo died. Michelangelo then lived on for another 45 years.[cxv]

Leonardo was first introduced to informal studies of Latin, geometry, mathematics and many other subjects as a 14-year-old apprentice in Verrocchio's studio. There, he, along with other interns, was exposed to drafting, chemistry, metallurgy, metal working, plaster casting, leather working and carpentry, as well as drawing, painting, sculpting and modeling. By 1472, at age 20, Leonardo was considered as a master in the Guild of Saint Luke, a respected organization of artists and doctors of medicine.[cxvi]

Leonardo conducted morgue studies which carefully recorded physiological effects of age and human emotions—rage in particular. He also dissected and studied anatomies of various animals including horses, cows, monkeys, bears, birds and frogs. Comparing their anatomies with humans, he concluded that the heart is central to the circulatory system in all. He demonstrated how the heart functions by creating wax models of the cerebral ventricles and a glass aorta to observe blood circulation through the aortic valve using grass seeds in water to watch flow patterns.

As with other great Italian painters of his time, Leonardo's works reflected a fresh, new natural world perception—fidelity, vigor, graceful human body representation and mastery of perspective. In addition to such masterpiece works as the *Mona Lisa, The Last Supper,* and his iconic drawing of the *Vitruvian Man,* many of the more than 13,000 pages of recorded drawings, notes and scientific thoughts on nature he produced throughout his life fuse art with natural philosophy—the forerunner of modern science.

In 1500, Leonardo, along with his assistant and a mathematician friend, fled Milan for Venice as guests of monks in the Monastery of Santissima Annunziata and St. John the Baptist. In 1502 he traveled throughout Italy in the service of Cesare Borgia, the son of Pope Alexander VI, as his chief military architect and engineer. In this role, he created a strategic map which described defensive methods to protect

Cesare Borgia's stronghold town Imola from naval attacks. Leonardo produced another defense map for the Chiana Valley which included a proposed sea dam to supply Florence with water throughout all seasons. Leonardo went to Florence in 1507 to sort out some estate problems with his brothers following the death of their father, then moved back to Milan in 1508. Between 1513 and 1516, he spent extended periods in the Vatican in Rome under Pope Leo X where Raphael and Michelangelo were also very active.

By this time, Leonardo had cultivated important royal and religious mentors. King Francis I of France recaptured Milan in 1515, and Leonardo was known to be present at a meeting of Francis and Pope Leo X in Bologna. Francis commissioned him to create a mechanical lion that could walk forward and then open its chest to reveal a cluster of lilies. In reward for his services, Francis awarded Leonardo a comfortable pension and prestigious manor house, now a public museum, near the king's residence at the royal Château d'Ambroise.

As historian Liana Bortolon wrote in 1967:

> *Because of the multiplicity of interests that spurred him to pursue every field of knowledge...Leonardo da Vinci can be considered quite rightly, to have been the universal genius par excellence, and with all disquieting overtones inherent in that term. Man is uncomfortable today, faced with a genius, as he was in the 16th Century, five centuries have passed, yet we still view Leonardo with awe.*[cxvii]

ADVANCEMENTS OF SCIENCE AND INDUSTRY

THE 16TH THROUGH 19th centuries marked a trend towards independent thinking that provided revolutionary scientific, technological and cultural foundations which are now generally taken for granted in contemporary society.

Much of the impetus for this new and irreversible era of proactive inquiry, free-thinking and innovation originated to a large extent in 16th century Western Europe during a "Scientific Revolution" which blended into an 18th century period of philosophical "Enlightenment." These developments, in turn, provided knowledge, methodologies and inventions that gave rise to a globally transformational "Industrial Revolution."

Early free thinkers whose "heretical" scientific inquiries and concepts challenged Roman Catholic Church and Protestant doctrines famously did so at great peril. This oppressive circumstance gradually diminished in combination with the disintegration of the Holy Roman Empire. This disintegration was influenced by growing reactions against

Church corruption and a century of crisis among divergent Christian beliefs that weakened intellectual conscriptions to religious orthodoxy.

A thirty-year religious war between Catholic and Protestant states (1618-1648) irrevocably changed the map of Europe.

The complex saga began as a battle between Catholic and Protestant states when future Holy Roman emperor Ferdinand II in his role as king of Bohemia attempted to impose Roman Catholic absolutism on his domains along with Austria. After Ferdinand triumphed in a five-year struggle, King Christian IV of Denmark saw an opportunity to gain valuable territory in Germany to balance his earlier loss of Baltic provinces to Sweden.

Christian's ultimate defeat in 1629 finished Denmark as a European power. After ending a four-year war with Poland, Sweden's Gustav II Adolph invaded Germany, winning over many of its princes to his anti-Roman Catholic, anti-imperial cause.

Poland, in turn, having been drawn in as a Baltic power coveted by Sweden, pushed its own ambitions by attacking Russia and establishing a dictatorship in Moscow under its future king Wladyslaw IV Vasa. A Russo-Polish Peace of Polyanov in 1634 ended the Poland tsarist throne, but freed Poland to resume hostilities against its Baltic archenemy. Sweden had become embroiled in German conflicts where three denominations vied for dominance: Roman Catholicism, Lutheranism and Calvinism.

Meanwhile, a network of Protestant town and principalities that relied on the anti-Catholic powers of Sweden and the United Netherlands threw off the yoke of control by Spain following an 80-year-long struggle. A parallel struggle involving the French rivalry with the Hapsburg Empire of Spain ended with France as the victor.

As the Thirty-Year War evolved, it became less about religion and more about which groups would ultimately govern Europe.

By 1648, the balance of power in Europe had dramatically changed. Spain had lost control not only the Netherlands, but also its dominance in Western Europe. France became a chief Western power, while Sweden controlled the Baltic. The United Netherlands became recognized as an independent republic, and member states of the Holy Roman Empire were granted full sovereignty.

The ancient notion of a Roman Catholic Empire of Europe headed spiritually by a pope as emperor was finally abandoned.

A Scientific Revolution

The aftermath of the Thirty-Years War more fully catalyzed a new scientific era which began a century earlier in which members of the educated class turned increasingly to direct observations and methodological empirical evidence to investigate natural phenomena.

A crowning achievement of this centuries-long period was great strides in better understanding natural laws which govern the dynamic architecture of our planet, Solar System and celestial neighborhood.[cxviii]

Many historians are inclined to mark the Scientific Revolution threshold beginning with Nicolaus Copernicus's heliocentric hypothesis which he published in 1543. His theory, one which influenced other important visionaries who followed, positioned the Sun near the center of the Universe, motionless, with the Earth and other planets orbiting around it in circular paths modified by epicycles and at uniform speeds.[cxix]

Soon after Copernicus's proposition confounded Church

teachings of an Earth-centered Universe, Danish astronomer Tycho Brahe (1546-1601) observed that with the exception of the Earth, all five planets, known at that time, revolved around the Sun. By holding to an Earth-centric belief, he evaded Church wrath.[cxx]

William Gilbert (1544-1603) undertook original studies which determined that our planet Earth somehow causes all magnetic phenomena in its environment. Applying experiments with observable proofs, Gilbert also distinguished differences between electricity and magnetism.

Johannes Kepler (1571-1630) validated Copernicus's radical heliocentric theory and postulated that planetary orbits are actually elliptical. Kepler recognized that the Sun controls planetary motion, and determined that the orbital periods of planets relate to their distances from the Sun.

Also following Gilbert's lead, Kepler explored the possibility that there might be some kind of a magnetic cosmic force that controls planetary positions and movements—the first known insight into gravitational theory. Kepler's empirical data and mathematics provided an essentially correct planetary picture which formed a fundamental basis for Isaac Newton's later seminal advances in cosmic science.

The same year that Copernicus published his planetary motion laws, Galileo Galilei (1564-1632) constructed a telescope to view planets and moons far better than ever before. Galileo discovered that Jupiter and its four moons move in concert around the Sun, and that Earth and its Moon revolve around the Sun as well. Observing phases of Venus, Galileo concluded that it also revolves around the Sun.

Often regarded as the "Father of Modern Physics," Galileo's 1610 *The Starry Messenger* publication changed concept of the Universe's structure beyond conventional imagination, one where even the Milky Way galaxy seemed to be part of a much larger Universe where stars extended outward in every direction.

In addition to supporting Copernican theory, Galileo showed that its principles could be characterized by mathematical formulations. For example, he demonstrated the inseparability of time and motion by a discovery that distances of fallen objects accelerate as the square of time. A body falls 16 feet in one second; therefore, it will fall 4 x 6 = 64 feet in two seconds. This proven formula refuted Aristotle's then-accepted common-sense contention that fall rate is proportional to weight.

While the pope initially showed some enthusiasm towards Galileo's cosmological discoveries, his ideas later sparked vitriolic criticisms from powerful Church and university critics. Frenzied fear of losing control over the faithful, his most dangerous notion that science took precedence over theology, resulted in being labeled a heretic which drew attention from the Inquisition tribunal.

One charge levied against Galileo was that he was unable to present quantitative evidence to support his radical heliocentric-rotating-Earth cosmology theory. English mathematician, physicist Sir Isaac Newton's (1642-1727), accomplished this proof in his 1687 book *Principia* which provided equations for gravity and for accurately predicting planetary motion. After determining Earth's orbital size, distances to nearby cosmic bodies could be calculated. Triangulation and stellar parallax then enabled mathematical distance determination.

Nevertheless, the Church snubbed Newton's clear proof of Galileo's correctness and held to their erroneous beliefs for five more centuries.[cxxi]

Although Cardinal Bellarmine banned Galileo's writings, they were smuggled north where the Western intellectual struggle resumed. Ironically, in 1930, Pope Pius XI canonized Bellarmine.[cxxii]

Observing that polished blocks move across a polished table more readily than on unpolished surfaces, Galileo predicted that movement of objects on frictionless surfaces proceeds forever. This concept of "inertia" was advanced by René Descartes at a very small scale, where: a particle at rest remains at rest without an impetus or push, while a particle in motion moves in a straight line at the same speed unless deflected by collision with another particle.

Descartes (1596-1650) perceived the Universe as a machine which is ordered by mathematical laws, an atomist concept wherein an infinite number of particles moved as imposed by God at their creation. He pondered that some unknown invisible force must hold planets in their orbits.

A strong Catholic, Descartes championed the concept of a scientific-religious "dualism"—wherein human spirituality and mechanical Earthly matter coexisted separately.

Descartes' reductionist philosophy advocated that the natural physical world can be best understood when analyzed from unique arrangements of its simplest parts. He believed that science embodied use of hypothesis and experiment to clearly characterize the nature of all things in terms of extension, size, shape, number, duration, specific gravity and relative position.[cxxiii]

In "Discourse on Method," Descartes famously concluded his own spiritual existence could not be doubted. He originally wrote: "je pense, donc je suis" (French), "cogito, ergo-sum" (translated into Latin), and later appearing in English, "I think, therefore I am."

Late 17th century scientist Robert Hooke (1635-1703), who is most broadly recognized for his micrographic discovery and naming cells, is less known for his provident theory that the same force governing planetary motion also applies to falling bodies.

Hooke's writings dating back to 1664 propose a concept of gravitational attraction which he describes as:

> ...such a Power, as causes Bodies of a similar or homogeneous nature to be moved one towards the other, till they are united... Planets are of the same nature as the sun and hence are attracted to it. Comets are not related, and they are repelled.[cxxiv]

Although Hooke made no direct reference to a centrifugal force providing "gravity," he recognized that a body revolving in orbit must be continually diverted from its inertial path by some force directed toward a center.

Hooke also stated his conviction that gravity decreases in power in proportion to the square of the distance, a view that may very well have influenced Newton who himself acknowledged in 1686 that correspondence with Hooke had stimulated him to demonstrate that an elliptical orbit around a central attracting body placed at one focus entails an inverse square force.

While having proposed the problem of the dynamics of elliptical orbits, Hooke also admitted his inability to solve it. That challenge was successfully addressed by Newton.

Isaac Newton's work, which synthesized Descartes' mechanistic philosophy, Kepler's laws of planetary motion and Galileo's laws of terrestrial motion explained this force. Reasoning that an apple falling from a tree demonstrated a force counteracting centrifugal force, Newton's mathematical analysis of this force, gravity, explained a great previous mystery.

Newton concluded that the Sun pulls planets with an attractive force that decreases inversely as the square of the distance from the Sun, and bodies falling toward the Earth conform to the same law. Newton also calculated elliptical planetary orbits and their speeds.

Newton's contributions to science are remarkable. He invented calculus before age 23. By 1684, he determined the mass and distance laws of gravity. By 1687, he formulated the three laws of motion and developed a method of quantifying centrifugal force.

Newton also determined that prisms break white light into component colors—different light refraction for different wavelengths. Then recognizing that chromatic aberrations in microscopes and telescopes relate to light refraction, the bending of light waves, he illustrated this phenomenon using a prism and then built a reflecting telescope corrected for chromatic aberration.[cxxv]

Whereas Newton and Aristotle believed that light travels instantaneously, in 1676 Danish astronomer Ole Romer showed that light has a finite speed. Observing that eclipses of Jupiter's several moons showed different timing than predicted depending upon variable distances separating Jupiter from Earth, Romer ascribed these discrepancies in light travel times.

Other great thinkers of the time were directing their attention to a variety of different scientific inquiries. Among these, the invention of the microscope opened up new worlds of understanding at a very tiny scale.

During the early 1600s, Galileo reportedly described an ability to focus his telescope to view small objects close up by looking through the "wrong" end. He subsequently improved the design of the device in 1624 based upon a compound instrument with a convex objective and a convex eyepiece (a "Keplerian" microscope) he had seen in London, perhaps one designed by Cornelius Drebble.

Dutch natural history student Anthony van Leeuwenhoek applied the microscope invention in 1650 to discover protozoa, rotifer, red blood corpuscles, capillaries and sperm cells in semen. The latter observation that a sperm combines with a female egg represented a radically new concept of reproduction which dispelled a mystical notion of spontaneous life generation.

In 1795, about a century and one-half later, English physician Edward Jenner invented the smallpox vaccination to save many lives from ravaging outbreaks of microscopic pestilence.[cxxvi]

A New "Age of Reason"

Dated generally between 1685 and 1815, the eighteenth century "Age of Reason" or "Enlightenment" was a movement that spread throughout Western Europe, England and the American colonies which dominated and radically reoriented intellectual, philosophical and political discourse.

Following closely on the heels of the Renaissance, The Enlightenment was, at its center, a celebration of ideas about what the human mind was capable of and what could be achieved through deliberate action and scientific methodology to advance individual opportunity and progress.

A strong new spirit of egalitarianism held the purpose of fair treatment of all people. Contending that freedom and democracy were fundamental rights of all people, not gifts bestowed upon them by beneficent monarchs or popes, growing numbers of voices openly expressed sharp criticism of Church oppression and obstruction of free inquiry.

The Enlightenment championed scientific and humanistic knowledge, a quest for secular understanding to further human rights, dignity and progress. Philosopher Bertrand Russell characterized the movement as a revaluation of independent intellectual activity—a clearing of religious thought to allow seeking of knowledge where darkness prevailed.[cxxvii]

French philosopher Denis Diderot (1713-1784) produced the signature publication of the period, *Encyclopedie*, which brought together leading authors to produce an ambitious compilation of knowledge. Diderot also wrote many of the articles himself, a strategy designed, in part, to avoid French censors.

Diderot's monumental compendium of then-current knowledge challenged Roman Catholic Church authority and also that the aristocratic French government, both of whom tried unsuccessfully to suppress it.

Diderot championed the value and uniqueness of the individual and promoted an optimistic belief that all knowledge could be acquired through scientific experimentation and the exercise of reason. He also advocated that education should be tailored to the abilities and interests of the individual student, and that students should learn to experiment and conduct research rather than simply acquire outside knowledge.

German philosopher Immanuel Kant (1724-1804), who worried that accepted knowledge had grown too dependent upon the thinking of just a few people, coined the Age of Reason motto: *"Dare to know! Have courage to use your own reason!"*[cxxviii]

Above all, Kant insisted that every rational being had both an innate right to freedom and a duty to enter into a civil condition governed by a social contract in order to realize and preserve that freedom.

Kant argued that the power of the state must be limited to protect citizens from the arbitrary exercise of authority, wherein the concept of "state" can be variously translated as the "legal state," "state of rights," or "constitutional state" in which the exercise of government power is constrained by law. This approach is based upon the supremacy of a country's written constitution—a supremacy which must create guarantees of a peaceful life as a basic condition for the happiness of its people and their prosperity.

Influential English philosopher John Locke (1632-1704) argued that human nature was mutable, and that knowledge must be gained through accumulated experience rather than by accessing some sort of outside truth.

Locke championed a principle that all people are equal and independent, with a natural right to defend their "life, health, liberty, or possessions."

Together with essays on religion, which provided an early model for the separation of church and state, Locke deeply influenced America's founding documents. Thomas Jefferson echoed Locke's concepts in the first sentence of the Declaration of Independence: "Human equality, and the right to life, liberty, and the pursuit of happiness." [cxxix]

Jean-Jacques Rousseau and Voltaire were prominent torchbearers of Enlightenment literature and philosophy.

French writer Rousseau (1712-1778) was a strong advocate for reform on behalf of social empowerment and democracy which remained influential long beyond his lifetime. His 1762 book *The Social Contract* argued against the idea that monarchs are divinely empowered to legislate, asserting instead that only the people are sovereign to hold all-powerful rights.

Rousseau concluded, stating, "Let us then admit that force does not create right, and that we are obliged to obey only legitimate powers." Here, the ability to coerce is not a legitimate state power, and there is no rightful duty to submit to it.

Voltaire (1694-1778) was, in fact, a pen name of Francois-Marie Arouet. He likely used this pseudonym device to shield him from persecution for pointedly barbed criticisms against the Roman Catholic Church which he reviled as intolerant, backward, and too steeped in dogma.

In keeping with many other Enlightenment thinkers of his era, Voltaire condemned injustice, clerical religious abuses, and while believing in a supreme being, regarded formalized religion as superstitious and irrational.

Voltaire vigorously emphasized empirical natural science that served in his mind as a necessary antidote to vain and fruitless philosophical investigations. Politically, he despised democracy as rule by mobs and believed that an enlightened monarchy informed by counsels of the wise was best suited to govern.

Although the political climate in the American colonies was vastly different than Europe, philosophies emerging from the Enlightenment had profound influences upon the New World as well. Among these colonist leaders, Benjamin Franklin and Thomas Paine—each in their own way—took up the rational thinking mantle.

For Paine (1737-1809) the new ideas in Europe likely prompted a desire to regard the colonies separate and independent from the British Crown. His *Common Sense*, an impassioned yet well-reasoned plea for independence, was instrumental in gathering supporters to this cause with the rallying cry of "No Taxation without Representation."

Benjamin Franklin (1706-1790) adopted a more utilitarian philosophy. While recognizing a need to become independent of the British Empire, he also foresaw the difficulties of forging a strong and lasting union out of disparate and competing colonial interests.

In 1757, Franklin was delegated to go to England as an agent of the Pennsylvania Assembly with the purported purpose of persuading the family of William Penn, as the proprietor of

Pennsylvania, to allow the Colonial Legislature to tax its un-granted lands. The mission's real aim, however, was to oust the family from power, and to make the colony a royal province.

Franklin then spent the next 18 years in London, influencing fellow colonists and British sovereigns alike to suspect disloyalties to both camps. His public persona as a "royalist" became reinforced through privileged political British connections which enabled him to have his son, then age 31, appointed Royal Governor of New Jersey.[cxxx]

As the eighteenth century drew to a close, passionate calls for social reform and a utopian, egalitarian society ebbed. Nevertheless, the world of Western and Colonial thought had been transformed. Science had been propelled forward by that time such that traditional authority of the Church was in real jeopardy. Monarchs no longer ruled by Divine Right, and common citizens had opened frank conversations and engagements influencing governance policies and the course of global events.[cxxxi]

If there was a historical moment that can be said to mark the beginning of the end of the Enlightenment, it was the French Revolution. France in 1789 had devolved into anarchy where sadism perpetrated by French citizens on each other was anything but enlightened.

The French Revolution led to the rise of Napoleon a decade later.

Enlightenment ultimately gave way to 19th century Romanticism when many poets and philosophers turned away from deductive science to emphasize knowledge and imagination gained through human intuition and emotion. Prevalent themes of Romantic literature included the celebration of nature and sublime beauty, the idealization of rural lifestyles, and the rejection of rationalism, social convention, organized religion and industrialization.

British romantic poet George Gordon "Lord" Byron (1788-1824) promoted defiance, rebellion, noble deeds and contempt for tradition. Although Americans Henry David Thoreau and Ralph Waldo Emerson rejected commercialism and championed personal spiritualism; they more readily accepted science than their European counterparts.[cxxxii]

Bertrand Russell referred to Romanticism as a "cult of emotions." Yet as history continues to demonstrate, in addition to great art and literature, emotional passions also drive revolutionary scientific and technological discoveries and developments.[cxxxiii]

As Wane Bundy observes in his book *Out of Chaos: Evolution from the Big Bang to Human Intellect*, a struggle between irrational and rational thinking may be an essential aspect of progressive civilization:

> *By indirection, irrationality may promote new, useful approaches to problems—finally reconciled by rationality. Ambivalence seems an innate condition of the human brain and the way of nature. By struggling with the extremes, our minds become informed and prompted toward the most workable solution, sometimes toward the strongest bias.*

Bundy concludes:

> *Perhaps the most prominent example is the struggle between religion and science.*[cxxxiv]

An Industrial Revolution

Wayne Bundy points to modern civilization through the Scientific Revolution and the Enlightenment to the present as a revival of Ancient Greek thought. Much appreciation also is owed to powerful lessons and new ways of thinking advanced by Copernicus, Galileo, Descartes and other great minds who awakened and enabled another radical revolution of human industry, a new age of machines.

Near the 18th century's end, a rapid transition from dependence upon small craft shops using hand-production methods to the establishment of new technologies, economies and lifestyles rapidly emerged in Europe and America. Generally dated from about 1760 to sometime around the mid-1800s, the early Industrial Revolution featured mechanization of textile production.

A second phase leading to an unprecedented rise in European population and economic growth began after about 1870. This "Second Industrial Revolution" featured new steel making processes, large-scale manufacture of precision machine tools and the use of increasingly advanced machinery in steam-powered factories.[cxxxv][cxxxvi]

The Industrial Revolution began with textile production in Great Britain, then the world's leading commercial nation, a global trading empire with substantial control over the North American colonies and the Caribbean. Britain also exercised significant trade influence over the Indian subcontinent through political ties with the powerful East India Company.[cxxxvii]

Following the early 16th century discovery of a trade route to India around Africa, the Dutch established the East India Company and other smaller companies to engage in trade throughout the Indian Ocean region and North Atlantic Europe. Cotton textiles purchased in Eastern India and sold in Southeast Asia comprised one of the largest segments of this commerce. Cloth represented more than three-quarters of all the East India Company's exports by the mid-1760s.[cxxxviii]

Sometime after 1000 AD, hand-manufactured cotton

textiles had already become a major trade industry in tropical and subtropical regions in parts of India, China, Central America, South America and the Middle East. Cotton cloth could be used as a medium of exchange almost everywhere.

Europe depended upon favorable growing conditions on southern colonial plantations for cotton imports. However, the raw material was costly due to difficulties in removing seeds, putting British textile producers at a trade disadvantage with Indian cloth-goods.

In 1794, U.S.-born inventor Eli Whitney (1765-1825) developed a revolutionary machine that radically changed this trade balance. His patented cotton gin dramatically sped up the process of seed removal by applying a combination of wire screen and small wire hooks to pull cotton fibers through the device as brushes continuously removed lint to prevent jams.

A person using the new cotton gin could remove as much seed in one day as previously required two months for an individual to hand-process. The device increased the productivity of removing seed from cotton by a factor of 50. As a result, cotton had become America's leading export to supply Europe's need for raw textile material by the mid-19th century.[cxxxix],[cxl]

The Industrial Revolution also mechanized the spinning and weaving of cloth which had traditionally been accomplished as a "cottage industry" principally for domestic consumption. Home-based workers produced goods under a "putting-out" contract with merchant sellers who typically provided the raw cotton materials.

Farmers' wives conventionally did the spinning off-season, while the men did the weaving. Using the spinning wheel, it took between four to eight spinners to supply one handloom weaver.[cxli]

A "flying shuttle," patented in 1733 by John Kay in

England, along with later improvements, doubled the output of the weaver. His invention also worsened the imbalance between spinning and weaving.

An early spinning breakthrough occurred in 1770 with British inventor James Hargreaves' patented the "spinning jenny." The device worked in a similar manner to the spinning wheel by first clamping down the fibers, then by drawing them out, followed by twisting. However, the spinning jenny produced a lightly twisted yarn only suitable for "weft" (the transverse thread drawn through and inserted over-and-under the longitudinal threads on a frame or loom—the "warp").

A spinning frame, or "water frame" patented in 1769 by Richard Arkwright was able to produce a hard, medium count thread suitable for warp, finally enabling mechanically assisted 100% cotton cloth to be made in Britain.

Samuel Compton's "spinning mule" introduced in 1779 yielded finer thread than hand-spinning, and at a much lower cost. This finally enabled Britain to produce highly competitive yarn in large quantities. The device combined features of the spinning jenny and water frame in which spindles were placed on a moving carriage. The system went through an operational sequence during which the rollers stopped while the carriage moved away from the drawing roller to finish drawing out the fibers as the spindles started spinning.

Entrepreneur Richard Arkwright brought ongoing advancing cloth production processes together in a mechanized cotton mill factory. Other inventors increased the efficiency of the individual steps of spinning (carding, twisting, spinning and rolling) so that the supply of yarn increased greatly.

Mechanized cotton spinning powered by steam or water increased the output of a worker by a factor of around 500. A power loom alone increased the output of a worker by a factor of over 40. And while large production gains also occurred in spinning and weaving of wool and linen, they were not nearly as great as those in cotton.[cxlii]

Although most of the power during the early mechanization period was supplied by water and wind, a rapid transition to steam energy occurred after 1800.

London inventor and entrepreneur Thomas Savery patented and constructed the first commercial steam power in 1698. The low-lift one-horsepower system combined a vacuum and pressure pump used in various water works and mine water-removal applications.

The first successful piston steam engine was introduced in Britain by Thomas Newcomen sometime before 1712. Its principal uses were to drain previously unworkable deep mines and to power municipal water supply pumps.

Fundamental steam engine improvements were introduced by Scotsman James Wat and business partner Englishman Matthew Boulton in 1778. Closure of the upper part of the steam cylinder redirected low-pressure steam to drive the top of pistons rather than venting it into the atmosphere as Savery and Newcomen had done. Use of a steam jacket and a separate steam condenser chamber also did away with the cooling water that had previously been injected directly into the cylinder, wasting steam in the process.

Evolutionary steam engine efficiency improvements resulted in enormous fuel savings, amounting to three-quarter or more reductions in coal use per horsepower-hour over Newcomen's.

Adaptation of stationary steam engines to rotary motion made them suitable for industrial uses. Key among these were for the development and mass production of precision machine tools, such as the engine lathe, planning, milling and shaping machines. Powering by these engines enabled all the metal parts to be easily and accurately cut, which in turn, made it possible to build larger and more powerful engines.

Improved power-to-weight ratios of new high-pressure steam engines made them suitable for mobile transportation applications. Providing lighter weight and smaller size for a given horsepower than stationary systems was accomplished by exhausting used steam directly into the atmosphere, thus doing away with a condenser and cooling water.[cxliii]

Widespread railroad development after 1800 made possible by steam engine advancements was also greatly enabled by major iron production innovations needed to create the many miles of tracks along with other essential products needed to support a growing population and industrial economy. Included was bar iron used as the raw material for making hardware goods such as nails, wire, hinges, horseshoes, wagon wheel rims, chains and structural shapes.

A small amount of bar iron was converted into steel. Most of the early cast iron was refined and converted to bar iron as well, although with substantial inefficiencies.

A major improvement in iron production efficiencies during the Industrial Revolution resulted from the replacement of wood with coal, which was more abundant and less expensive. Coal required much less labor to mine than that involved in cutting wood and converting it to charcoal. In addition, other applications such as construction were causing wood to become increasingly scarce.

Another factor limiting the iron industry prior to the Industrial Revolution was a scarcity of waterpower to power blast bellows. In addition, the leather used in those bellows was expensive to replace.[cxliv]

Iron master John Wilkinson patented a high-pressure hydraulic-powered blowing machine to blast air in 1757 that solved both problems. The design was later improved by making it double-acting, which allowed higher blast furnace temperatures.

The substitution of coke (conversion of coal by heating it in the absence of air) for charcoal greatly lowered the fuel cost of crude pig iron and wrought iron production. Using coke enabled economies of scale afforded by these larger blast furnaces.[cxlv],[cxlvi]

Henry Cort developed two significant iron manufacturing advancements in, rolling (1783) and puddling (1784). The rolling mill was 15 times faster than hammering wrought iron.

The puddling process produced a structural grade iron at a lower cost than forging by means of decarburizing molten pig iron by slow oxidation in a furnace. This remained extremely hot, backbreaking work which involved manually stirring the material with a long rod. Few puddlers reportedly lived to reach the age of 40.

Hot blast patented by James Beaumont Neilson in 1828 greatly increased fuel efficiency in iron production in the following decades. Attributed by some as one of the most important developments of the 19th century, it saved energy in making pig iron by using preheated combustion air, reducing fuel consumption by one-third using coke or by two-thirds using coal.[cxlvii]

Steel was an expensive commodity prior to the Industrial Revolution. Accordingly, it was used only where iron would not do, such as for cutting edge tools and for springs. A crucible technique developed by Benjamin Huntsman in the 1740s enabled large-scale production of cheaper iron and steel which aided a number of industries. Included were commodities such as nails, hinges, wire and other hardware items.

Machines as Companions and Competitors

Perhaps most impactful, new power and metallurgical advancements enabled the development and mass production of precision machines.

Pre-industrial machinery was built by craftsmen—millwrights built water and windmills, carpenters made wooden framing and smiths made metal parts. As the Industrial Revolution progressed, ever cheaper and more precise machine-made tools and metal parts became increasingly common.

The first machine tools included the screw-cutting lathe, cylinder boring machine and the milling machine. These led to capabilities enabling the economical manufacture of large numbers of precision threaded metal fasteners such as screws, bolts and nuts.

In the 1770s, Henry Maudslay built a lathe which could cut machine screws of different thread pitches. These were the first machines for mass production capable of making components with a high degree of interchangeability.

The concept of interchangeable parts first took ground in the firearms industry when French gunsmith Honoré LeBlanc promoted the idea of using standardized gun parts. Before this, individual firearms were made by hand and varied slightly from one to another. Thus, each weapon was unique and could not be easily fixed if broken.

It wasn't until cotton gin inventor Eli Whitney introduced the idea in the United States Department of War in the 19th century that the development of interchangeable parts for small firearms really took off. Whitney had trained a large unskilled workforce using standardized equipment to produce large numbers of identically replaceable gun parts at a low cost and within a short amount of time.[cxlviii]

In the half century following the invention of the fundamental machine tools, the industry became the largest value-added industrial sector of the US economy.

In 1901, Ransom Olds created and patented the assembly

line, a factory process which allowed his car manufacturing company to increase output by 500 percent in one year. A Curved Dash model was able to be produced at what then was an exceptionally high rate of 20 units per day.

Henry Ford improved upon Olds' assembly line concept by using the moving platforms of a conveyor system. The vehicle chassis was towed by a rope that moved it from station to station, allowing a progressive sequence of stationed workers to assemble each part.

Ford's revolutionary assembly method enabled a "Model T" to be produced every ninety minutes, totaling nearly two million units in one of their best years. Often credited as the father of the assembly line, Ford would be more appropriately characterized as the father of automotive mass production.

Life Quality Contributions and Consequences

Some economists, such as Robert E. Lucas, Jr., say that the real impact of the Industrial Revolution was that "for the first time in history, the living standards of the masses of ordinary people have begun to undergo sustained growth...Nothing remotely like this economic behavior is mentioned by the classical economists, even as a theoretical possibility." [cxlix]

The Industrial Revolution was the first period in history during which there was a simultaneous increase in both population and per capita income.

During the Industrial Revolution, life expectancy increased dramatically. The percentage of children born in London who died before the age of five decreased from 74.5% in 1730—to 31.8% in 1810-1829.[cl]

Until about 1750, in part due to malnutrition, life expectancy in France was about 35 years and about 40 years in

Britain. The U.S. population at the time was adequately fed, much taller on average and had a life expectancy of 45-50 years.

A very major contribution of the Industrial Revolution was food abundance, essential to nourish growing populations. The increase in food abundance over the past 200 years constitutes what can legitimately be termed a Second Agricultural Revolution.

As Yuval Noah Harari points out:

> *Machines such as tractors began to undertake tasks that were previously performed by muscle power, or not performed at all. Fields and animals became vastly more productive thanks to artificial fertilizers, industrial insecticides and an arsenal of hormones and medications. Refrigerators, ships and airplanes have made it possible to store produce for months, and transport it quickly and cheaply to the other side of the world. Europeans began to dine on fresh Argentine beef and Japanese sushi.*[cli]

Harari and others also appropriately argue that while the growth of the economy's overall productive powers was unprecedented during the Industrial Revolution, living standards for many workers were very low. Histories of these early times bring to mind prevalent images of urban landscapes dominated by smoking chimneys and the sad plight of exploited coal miners sweating in the bowels of the earth.

Living standards and health gradually yet dramatically improved during the 19th and 20th centuries. Labor laws, for example, addressed exploitive working conditions and compensation practices and public health acts regulated industrial sewage disposal.

And while new and more efficient machines and processes of the Industrial Revolution yielded an explosion in human productivity, for a great many—craft and farm workers in particular—it cost them their jobs and livelihoods.

The mechanization movement started first with British lace and hosiery workers, then rapidly spread to other areas of the textile industry.

In 1811, angry mobs of newly unemployed weavers and other workers turned their animosity towards attacking machinery and factories that had taken their jobs. Riots by self-identified Luddites, supposedly followers of Ned Ludd, a mythical folklore figure, often turned violent. Many were arrested by British militia troops hired to protect industry and tried and jailed. Some were even hanged.

Unrest also occurred in other mechanized industry sectors. In the 1830s, for example, agricultural laborers in southern Britain destroyed threshing machines and burned hay bales.

Nevertheless, despite employment disruptions and workforce shifts, the Industrial Revolution created far more jobs than casualties. The abundance of more affordable products in combination with rapid increases in general consumer prosperity gave birth to a new capitalistic era of entrepreneurship, innovation and global commerce.

Liberalization of trade from an expanding merchant base allowed Britain to produce and use emerging scientific and technological developments more effectively than countries with stronger monarchies, particularly China and Russia.

Philosopher Karl Marx had predicted that capitalism would be overthrown by communism so that oppressed workers would finally be free. History didn't turn out that way.

Karl Marx got it exactly backwards.

New Wonders and Worlds of Discovery

Scientific and technological progress has advanced at an ever-accelerating pace, where inventions continuously spirit, enable and ultimately multiply new innovations and knowledge exponentially.

In the remarkably short span of a century, humans learned to harness lightning, to develop wings and to split atoms. Such discoveries and advancements have at once and forever transformed society in two fundamental ways. Just as they continue to represent exciting forces of promise, those same forces empower terrifying weapons of war.[clii]

A fast-paced 19th-20th century era of scientific discovery and technological progress actually began in 18th century BC Greece. From there it was rediscovered and rekindled during the 14th-17th century Renaissance, was catalyzed and objectified during the 17th-18th century Scientific Revolution, was culturally inspired during the 18th century Enlightenment, and was accelerated to warp-speed during the 18th-19th century Industrial Revolution.

The 18th-century invention of the steam engine rapidly transformed industries, railroads and later, along with the electricity-generating dynamo, electrified a whole new world of work and life-changing possibilities.

In 1864 James Maxwell developed a revolutionary theory of electromagnetic waves, and Marconi invented wireless telegraphy in the late 1890s. At about that same time, Wilhelm Conrad Röntgen discovered X-rays, French physicist Becquerel discovered radioactivity, Marie and Pierre Curie carried out their pioneering work on radioactivity using radium and Ernest Rutherford formulated an atomic structure theory which first described a nucleus encircled by electrons.

Electrifying Society

Electrification enabled by the invention of electromagnetic generators—powered by innovations of industrial-scale steam turbines—must certainly be credited as one of the top-transformative 20th-century developments. The original operating principle, now known as Faraday's Law, was discovered by Michael Faraday in the years 1831-1832, namely that an electromagnetic force is generated in an electrical conductor which encircles a varying magnetic flux. Faraday's first electromagnetic generator used a copper disc rotating between poles of a horseshoe magnet to produce a small DC voltage.

The first electric generator capable of delivering commercially practical power was the dynamo used in 1844 for electroplating. Modern dynamos capable of producing industrial-scale electricity were invented independently two decades later by Sir Charles Wheatstone, Werner von Siemens and Samuel Alfred Varley between 1866 and 1867. Siemen's design, which incorporated electromagnets rather than permanent magnets, greatly increased the dynamo output required for high power-demand applications such as electric arc furnaces used in the production of metals.[cliii]

The first power stations supplied direct current (DC) which was well-suited for a number of applications such as electric street railways, machine tools and certain industrial applications where speed control was important. Serbian-American electrical engineer Nikola Tesla's invention of alternating current (AC) soon became the option of choice for general electrification because it could be transformed to high voltages with low power losses and also enabled motors to run at very constant speeds.

Although Tesla had attended the Austrian Polytechnic in

Graz, Syria, on a scholarship, he left after his second year 1881 to work as a low-wage draftsman at the Central Telegraph Office in Budapest, Hungary, where he was soon promoted to a chief electrician position. Two years later Tesla moved to America and was employed by Edison's Machine Works to develop a high voltage arc lamp-based lighting system.

Recognizing that Edison's DC technology was incompatible with high-voltage requirements, Tesla proposed a revolutionary AC alternative which was rejected. He then left Edison's company in 1885 after only six months, and together with some investors, founded the Tesla Electric Light & Manufacturing Company. After proceeding to patent a new arc lighting concept along with new types of AC motors and electrical transmission equipment, the enterprise folded. Tesla lost control of the patents, leaving him broke.

In 1887, together with two new investors, Tesla then formed the Tesla Electric Company. The new company developed an AC induction motor, a concept affording large advantages for long-distance, high-voltage transmission. Engineers at Westinghouse Electric & Manufacturing recognized the importance of the design, and the company negotiated a licensing deal. A Westinghouse - General Electric merger arrangement later purchased the patent from Tesla's company.

Tesla's achievements following that period were indeed transformative. He accomplished the first successful wireless energy transfer to power electronic devices in 1891, conducted the earliest demonstration of fluorescent lighting, and influenced the development of modern electrical generators and turbine designs.

In 1893 the Westinghouse Electric Company implemented Tesla's AC system to light the World Columbian Exposition in Chicago. The demonstration proved to be more efficient than the direct current system marketed by Edison, and rapidly became the basis for most modern electric power distribution systems. In 1895, Tesla and Westinghouse developed the world's first hydroelectric power plant at Niagara Falls.

At the turn of the century, Tesla set up a laboratory in Shoreham, Long Island featuring a "Wardenclyffe Tower" project intended to provide intercontinental wireless communications as a more powerful transmitter in competition with a Marconi radio-based system which Tesla regarded as a copy of his design.

Tesla's investors dropped out after Marconi's system won out in December 1901 by successfully transmitting the letter "S" from England to Newfoundland. He died virtually penniless following an unsuccessful attempt to sue Marconi for infringement on his wireless patents.

Thomas Alva Edison, Tesla's earlier rival in the "electric current war," is recognized as one of America's most prolific inventors. His more than 1,000 patents include such innovations as incandescent electric lights, the microphone, telephone receiver, stock ticker, phonograph, movies and office copiers.

At age 20, Edison secured work in Cincinnati, Louisville, Indianapolis, Memphis and Boston as an itinerant Western Union telegraph operator. The job suited his interest in learning more and more about telegraphy, including how to improve the equipment.[cliv]

By 1969, Edison's entrepreneurship as an inventor began to really take off. His patent applications included a telegraphic stock ticker which became standard office equipment in America and Europe, and a printing telegraph for gold bullion and foreign exchange dealers. He also figured out how a central telegraph office could control the performance of equipment from remote locations and developed a method to transmit as many as four messages over a single wire.[clv]

On July 18, 1877, as Edison tested an automatic telegraph which had a stylus to read coded indentations on strips of paper, the friction revealed an unexpected hum that attracted his attention. As Douglas Tarr at the Edison National Historical Site in West Orange, New Jersey reported:

> *Edison seemed to reason that if a stylus going through indentations could produce a sound unintentionally, then it could produce a sound intentionally, in which case he should be able to reproduce the human voice...A talking machine!*[clvi]

Edison worked on and off over more than two decades to advance that concept to do much more than just talk. His innovation ultimately produced sound quality that brought high fidelity music to homes of world audiences.

In 1879, Edison's Menlo Park laboratory demonstrated the first high-resistance incandescent light which passed electricity through a thin platinum filament in a glass vacuum bulb to delay melting. After the original model worked only for an hour or two, Edison went on to try carbonized filaments made of almost every imaginable plant material, including some specially ordered fibers from the tropics. The best performer proved to be carbonized filaments of common cotton.[clvii]

Many more innovations followed during the late 1880s and early 1890s. In the area of photographic optics, for example, Edison demonstrated the potential of using tough, flexible celluloid motion picture film, worked out mechanical problems of advancing the film steadily across a photographic projection lens without tearing and linked a new motion picture camera with an improved phonograph featuring synchronized sound, producing the "Kinetoscope" that projected "talking" images on screens.

Edison's legacy of achievement is commemorated by numerous companies that bear his name. Included are: Edison General Electric (which merged with the Thomson-Houston electric company to form General Electric); Commonwealth Edison (now part of Exelon); Consolidated Edison; Edison International; Detroit Edison (a unit of DTE Energy); the Edison Electric Institute (a trade association); the Edison Ore-mining Company; the Edison Portland Cement Company; Ohio Edison (which merged with Centerior in 1997 to form First Energy); and Southern California Edison.

Splitting Light and Atoms

It's difficult to imagine anyone who exemplifies a greater genius in the popular minds of most people than Albert Einstein. The products of his thinking delivered far more than he originally advertised in a 1905 letter to his friend Conrad Habicht.

Einstein, then working as a low-level patent examiner, wrote:

> *I promise you four papers. The first deals with radiation and the energy properties of light and is very revolutionary, as you will see if you send me your work first.*

That paper postulated that light could be regarded both as a wave as well as a stream of tiny particle packages called "quanta."

Einstein went on to say:

> *The second paper is a determination of the*

> *true sizes of atoms...The third proves that bodies on the order of magnitude 1/1000 mm, suspended in liquids, must already perform an observable random motion that is produced by thermal motion. Such movement of suspended bodies has actually been observed by physiologists who call it Brownian motion.*

Using statistical analysis of random collisions, that third paper established the true existence of atoms and molecules.

Einstein continued that:

> *The fourth paper is only a rough draft at this point, and is an electrodynamics of moving bodies which employs a modification of the theory of space and time.*

This later became famously known as the "Special Theory of Relativity."

That same year, he was also working on a short addendum to that fourth paper which drew a relationship between energy and mass. The addition envisioned bending of light beams and warping of space. That relationship is briefly and most famously of all summarized as $E=mc^2$. His predictions of how much gravity actually bends light were later validated during a 1919 solar eclipse.

In 1895, 16-year-old Einstein imagined what it would be like to ride alongside a light beam. A decade later, this boyhood musing provided the conceptual foundation for two great advances of 20th-century physics: relativity and quantum theory.

Then in 1915, only one more decade after that, he

followed that light beam of imagination to produce his everlasting crowning scientific accomplishment. That General Theory of Relativity explained how "space-time" is warped by an interplay between matter, motion and energy.

Einstein likened this circumstance to rolling a bowling ball onto the two-dimensional surface of a trampoline. Then when some billiard balls are added, they move toward the bowling ball not because it exerts some mysterious attraction, but rather, because of the way it curves the trampoline fabric.

Here, space and time are not two separate things, but together form space-time where energy and mass are actually different forms of the same thing. How these mass versus energy determinations are measured is influenced by how fast the object and observer are moving relative to one another.[clviii]

A February 2016 announcement which the Royal Swedish Academy accurately described as "a discovery that shook the world" affirmed that Einstein had been proven right. Just as his 1916 General Theory of Relativity had predicted, sensitive Earth-based instruments recorded that gravity waves emanating from the collision of two black holes a billion light years away jiggled space-time with invisible cataclysms which reached us.

That faint "chirp" signal which was received at separate facilities in different states lasting only a fifth of a second was greeted by thousands of scientists as a loud opening bell for a whole new era of astronomical revelations.

Instruments at the U.S. Laser Interferometer Gravitational Observatory (LIGO) detected ripples in the space-time grid produced by a different type of event on August 17, 2017, which recorded the collision of two neutron stars. In addition to gravity waves, the spectacle released visible light which was observed by Earth-based telescopes. Initially appearing as a bright explosion of blue, the color soon faded to a deep red.

The discovery confirmed, as expected, that collisions of

neutron stars produce enormous gamma-ray bursts, along with about half of all heavy elements which are dispersed in gases that eventually settle down and condense to form new stars and planets.

Einstein was intellectually absorbed with a "wave-particle-paradox" whereby light can be measured either as "waves" of light or as energy "particles" depending upon which equipment we select to observe it.

It should be noted that those "waves" of light can also be measured as energy "particles" which don't contain any physical "stuff." Depending upon which equipment we select to observe it, some experiments show that light is wave-like, while others show that it is a particle-like phenomenon.

Thomas Young's 1903 experiments showed that light must be wave-like, while Einstein "proved" that it is particle-like.

Einstein's theory proposed that light is comprised of tiny particles (photons) analogous to a stream of bullets, whereby energy itself, is quantized. He termed this a "photoelectric effect."

Max Planck, the first physicist to calculate the sizes of "energy packets" (quanta) in various waves of light frequency (color) using his mathematical invention famously known as "Planck's constant." All of those packets of color, red for example, have the same size.

As Planck described Einstein's theory:

> ...the photons (the 'drops' of energy) do not grow smaller as the energy of the ray grows less; what happens is that their magnitude remains unchanged and they follow each other at greater intervals.[clix]

Einstein was not able to dispute the contradiction between light as a wave versus light as quanta, but simply took the contradiction as something which would probably be understood later. Nevertheless, while he is far more famous for two revolutionary theories of relativity, both were based upon his discoveries regarding the quantum nature of light which earned him a Nobel Prize.

Although Einstein never embraced what came to be recognized as a scientifically well-established yet counterintuitive "quantum mechanics" theory he is credited with advancing, neither could he dispute that it invariably "worked." For example, as discussed later in this book, principles of quantum theory are now being applied to create advanced computers with astounding processing capacities.

Niels Bohr, a Danish physicist who earned a Nobel Prize in 1922 for his contributions to quantum mechanics, argued famously with Einstein on the subject. Einstein lamented, "Alas, our theory is too poor for experience." Whereas Bohr replied, "No, no! Experience is too rich for our theory." [clx]

While Newtonian physics works wonderfully well to describe and predict events in our "everyday world," it cannot account for phenomena in the subatomic realm which appear to be governed by very different rules.

Just as Einstein's breakthrough Special Theory of Relativity affirms, appearances of observed subatomic events (such as light effects) are relative and dependent upon the observers. Since atoms are far too small to actually "see," all that scientists can do is speculate about what is there based upon certain observations regarding how atoms appear to behave.

Quantum mechanics takes this condition one very bizarre step farther. The very fact of being observed, and by whom, influences the very event being witnessed.

The new quantum theory model presents a vision of a subatomic world comprised of unimaginably small "particles" which have no material substance, yet for convenience, are statistically measured as quanta in terms of energy units in the same way as particles. These quanta unceasingly change measurable appearances from energy to mass and back, although "within a common identity."

Considering size distance comparisons between atoms and subatomic particles versus between our Solar System and planets, for example, distances between an atomic nucleus and its electrons are far greater.[clxi]

As described by science writer Gary Zukav, the difference between the atomic level and subatomic level is as great as the difference between the atomic level and the entire planet.[clxii]

For another comparison, Zukav asks us to imagine an atom as the size of a grain of sand in the center of the dome of Saint Peter's Basilica in the Vatican, with electrons the size of dust particles revolving around its outer edge. However, unlike dust particles which can be visualized as "things," quantum mechanics views subatomic particles only as "tendencies to exist" or "tendencies to happen" which can only be "seen" in the form of mathematical probabilities.

In addition to revolutionary contributions to sciences at all scales—ranging from the Universe to quantum subatomic quanta—Einstein's work also led to many important technological advancements by others. Included are photoelectric cells, lasers, fiber optics, semiconductors and nuclear power.

Regarding the latter, his discovery that $E=mc^2$ (where energy is proportional to mass multiplied by the extraordinarily huge number of the speed of light squared) is one of the most consequential scientific game-changers in human history. While that equation appears to be remarkably short and simple, it has since enabled humanity to harness the power contained in tiny atoms both to power prosperity and to annihilate itself.

Humanity Takes Flight

Dreams of flight likely date back to humankind's earliest conscious fascination with the soaring freedom of birds. During the mid-1400s, Leonardo da Vinci studied the structures and workings of their wings in attempts to produce machines that might bring such fantasies to fruition through a variety of mechanical devices. One of these—his previously mentioned hypothetical, un-tested flapping-wing "ornithopter"—was unsuccessfully attempted by many other inventors over the next four centuries.

The 18th-century discovery of hydrogen gas led to the invention of tethered and free-flying balloons which were first used for military surveillance purposes, these, in turn, led to the passenger-carrying rigid dirigible balloons pioneered by Ferdinand von Zeppelin in Germany, also referred to as "airships" which dominated long-distance flight until the 1930s.

The catastrophic ignition of hydrogen tanks used on the German Luftschiff Zeppelin company's longest-class dirigible, the LZ 129 Hindenburg, marked the beginning of the end of the airship's popularity. The May 6, 1937, disaster which killed 36 people at the end of its first North American transatlantic flight at the Lakehurst Naval Air Station in New Jersey had been preceded by crashes of several others, three of which cost even greater numbers of fatalities.

Although non-flammable helium was known to be the safest gas for airships, it was rare, and therefore far more expensive than hydrogen. The U.S. Government issued a Helium Control Act of 1927 to ban its export, virtually forcing the use of hydrogen for large-scale lighter-than-air passenger craft.[clxiii]

Late 19th-century experiments with heavier-than-air craft and early-20th century experiments in engine and aerodynamic technology innovations provided revolutionary foundations for modern aviation.

In 1891, American astronomer Samuel Pierpont Langley published a paper titled *Experiments in Aerodynamics*, and on May 6, 1896, launched the first two sustained-flight demonstrations of an unpiloted heavier-than-air craft which he launched by a spring-actuated catapult mounted on top of a houseboat on the Potomac River near Quantico, Virginia. The longest flight of these two on that day traveled 3,300 feet at about 25 miles-per-hour at top speed.

Langley launched another successful unpiloted demonstration witnessed by Alexander Graham Bell on November 28, 1896, which traveled nearly one mile. This was followed by a quarter-scale passenger engine-powered concept version he tested in 1901 and 1903.

Sadly, for Langley, his efforts to create the first engine-powered passenger-carrying aircraft ended nine days after a second abortive attempt on December 8, 1903. The Wright brothers accomplished this feat on December 17, 1903.

Orville and Wilbur Wright had built and tested a series of kite and glider designs prior to attempting to build a powered design. After the first partial-scale glider they designed flew poorly, they built a makeshift wind tunnel to test 200 wing designs to develop a superior full-size version.

The Wrights invented an innovative wing warping concept along with a steerable rear rudder for controlled flight, along with a low-powered internal combustion engine and specially shaped wooden propellers for optimum power efficiency.

Orville's historic 12-second "Flyer I" flight, which took place four miles south of Kitty Hawk, North Carolina, traveled a total of 120 feet. This was followed by one flown 852 feet by his brother Wilbur that same day which lasted nearly a minute.

The brothers continued to improve and test their designs at Huffman Prairie near Dayton, Ohio. Their third version became the first practical aircraft to fly consistently under full pilot control from its starting point safely and without damage. Wilber successfully piloted a Flyer III a record-breaking 24 miles in 39 minutes, 29 seconds on October 5, 1905.

Rocketing to New Heights

The space age was founded and shaped upon bold ideas and dedication of many great minds whose contributions of purpose, passion, professionalism and persistence have brought humankind to our present crossroads of great possibilities and uncertainties. Four among countless others of these visionaries include a remarkably innovative Russian school teacher, another Russian who survived terrible deprivations in a prison work camp, a German World War II rocket developer and an American who dared to believe that rockets can operate in a space vacuum.

Their combined story, and those who joined and followed them, is one of historic achievements, events and lessons born of triumphs and tragedies. It reveals a nexus of politically-manipulated and ideologically-shifting public rivalries between nationalistic pride and paranoia where space exploration and technology manifests full dimensions of civilizations' boldest dreams and greatest fears. In all cases, it has inexorably changed and expanded our world.

As historian Walter A. McDougal has observed in his 1985 book *Heavens and the Earth*, there is probably no more exemplary and ironic time and place to begin this narrative saga than in early Bolshevik Russia:

Larry Bell

Modern rocketry and social revolution grew up together in tsarist Russia. There is no anomaly in the fact that the most 'backward' of the Great Powers before World War I was the one that fostered violent rebellion against the chains of human authority and the chains of nature.

Russian thinkers dating back to the 1880s including Viktor Sokolsky have contemplated general possibilities of creating liquid-fueled rockets which are commonplace today. We can thank the writings of a self-taught high school mathematics and physics teacher in the small town of Kaluga south of Moscow for the concepts and calculations upon which such realities depend.

Broadly considered to be the "Father of Space Travel," Konstantin Eduardovich Tsiolkovsky (1857-1935) originally preoccupied his early years with personal design studies related to research into stellar radiation and design concepts for steam engines and metal-fabricated dirigibles. Then, upon conceptualizing possibilities for reaction-driven devices later called "rockets," he published an amazing book in 1883 titled *Free Space* which proposed a comprehensive, detailed and ingenious design for a liquid-fueled propulsion device for use in the vacuum of space...

...which, when combined chemically would yield per unit mass of resultant product such an enormous amount of energy.[clxiv]

Konstantin Tsiolkovsky's prolific productivity during the 1920s through early 30s was amazing, conceiving ideas for "a reaction engine" (1927-28); "a new airplane" (1928); "a jet-propelled

164

aeroplane" (1929); "the theory of the jet engine" (1930-34); "the maximum speed of a rocket" (1931-33); and the final classic work before his death, "space rocket trains" (1924-1934).

He told a group of students at the Zhukovsky Academy in 1934:

> *...I am not at all sure, of course, that my 'space rocket train' will be appreciated and accepted readily, at this time. For it is a new conception reaching far beyond the present ability of man to make such things. However, time ripens everything; therefore I am hopeful that some of you will see a space train in action.*

And they did. For example, Tsiolkovsky providently conceived "space rocket trains" which are now the standard multi-staging technique used to deliver payload elements to Earth orbits and planets. His enormous conceptual achievements led to the USSR's first "Mir" orbital space station which was realized in large part through the efforts of another important Russian designer in 1971.

Tsiolkovsky's rich legacy of design contributions guided design practices of a great officially unnamed Soviet engineer known by his colleagues as "SP" who led efforts which produced the USSR's first ICBM and launched the Space Age with the first orbiting satellite (Sputnik), the first dog, first man, first two men, first woman, first three men, first spy, communication satellites, the vehicles and spacecraft that first reached the Moon and Venus and passed by Mars and the Mir orbital station.

The Chief Designer's identity was concealed as a state security secret under orders from Stalin, Khrushchev and

Brezhnev until his death in 1966 during the peak period of a USSR race to beat America in landing its citizens on the Moon. Only then was he publicly honored as the "Hero of Socialist Labour."

Sergei Pavlovich Korolev's remarkable career began with a childhood passion for aviation.

By 1929 he developed his own unpowered glider. This primary flying interest soon turned to possibilities of using rockets to improve aircraft performance.

While still a young university student, Korolev joined a "Group for Studying Rocket Propulsion" (GIRD) and began working on a small gasoline-fueled propulsion engine-powered rocket weighing 40 pounds. Launched in August 1933, it flew for a total of only 18 seconds. Nevertheless, he enthusiastically predicted in an article titled *Towards the Rocketplane:*

> *Jet flight vehicles can develop flight speeds of 3,600 km/hr…and [can attain] immense altitudes [but that] practical resolution of this huge problem requires years of persistent work.*

GIRD activities by Korolev and his coworkers drew the attention of the Russian military. His rocket group became merged into a new government-headed organization called the "Reaction Propulsion Institute" (RNII) for key purposes of creating reliable and accurate rocket guidance and control systems.

A massive purge of USSR scientists, engineers and military leaders accused of trumped-up charges of spying for Germans led to the arrests of many RINN engineers and the execution of the organization's main sponsor, renowned military hero Marshal Tukhachevsky. On the early morning of June 27, 1938, two KGB agents took then-31-year-old Korolev into custody with no time to say goodbye to his three-year-old daughter Natasha.[clxv]

Having risen to a high-level RNII position, Korolev was taken to Lefortovo prison where he was interrogated and beaten. Upon asking for a glass of water he was hit on the head by a jug handed to him and called an "enemy of the people." He was then told, "Today is your trial" and was led down a long corridor into a room.

When the door opened, Kliment Voroshilov, one of Stalin's closest associates, entered. Imagining that Voroshilov would straighten out the problem Korolev told him *"I didn't commit any crime."* Voroshilov then shouted, "None of you swine ["svolochi" in Russian] have committed a crime. Ten years hard labour. Go! Next!" [clxvi]

Korolev was accused of collaborating with an anti-Soviet organization in Germany in order to "subvert a new field of technology." This event may not have been entirely unexpected following the arrest and prison sentencing of Valentin Glushko, another leading Soviet rocket designer three months earlier.[clxvii]

As with Glushko, Korolev received no trial. He was beaten and forced to confess. After receiving a 10-year sentence and having his family's property confiscated, he was moved from one prison to another. In October 1939, he was transferred to the most dreaded of all, the Kolyma forced labor camp in far eastern Siberia made infamous in the West in Aleksandr Solzhenitsyn's publication of *The Gulag Archipelago.*

Several thousand prisoners at Kolyma reportedly died from malnutrition, lack of shelter and harsh discipline each month...as many as 30 percent per year. Korolev's five-month-winter ordeal cutting trees, digging and pushing wheelbarrows at a Kolyma gold mine resulted in heart damage, a broken jaw and the loss of all teeth.[clxviii]

Following physically and emotionally grueling Kolyma experiences and numerous failed appeals, Korolev's case was reinvestigated and his sentence was reduced from 10 to 8 years in 1939 when Lavrentiy Beria was replaced by Nikolai Yezhov as Minister of Internal Affairs. He was then moved to greatly improved conditions at a penal institution known as Central Design Bureau 29 where most occupants were intellectuals, including scientists and engineers, and put to work in charge of wing design for a light bomber.

Korolev was later relocated to another penal institution in Kazan, Siberia, when Germans approached Moscow. Although technically freed in 1942, he voluntarily elected to stay in order to continue work he considered important and he soon became a chief designer for aircraft engines.

Then in 1944, he was moved once again to a penal facility in the Caucasus, where he once again began working as a rocket engineer until his formal release as a prisoner.

Korolev's early death at age 59 might be attributed in part to health problems arising from brutal imprisonment conditions suffered as a victim of the oppressive Stalin regime. Yet he was never known to speak to anyone about his hard treatment and privations until later, just a few days following a 59th birthday party. Late that night after other guests had departed, he confided a sad account to the world's first Earth-orbiting human Yuri Gagarin.[clxix]

By the late 1930s, when some ballistic missile development was occurring at the U.S. Jet Propulsion Laboratory, the Germans were already developing plans for a major Peenemunde Army Research Facility for fearsome V-2 rocket production in a small town located on Usedom Island on their northern coast. Those activities would ultimately be directed by a charismatic and effective engineer...Wernher von Braun.

Wernher had several characteristics in common with his Soviet counterpart, Chief Designer Korolev. As also with American rocketry pioneer Robert Goddard, all began their careers experimenting as rocket amateurs. In addition, although both von Braun and Korolev maintained spaceflight to the Moon and planets as key goals, they received early funding for military missile development, and although terms of punishment were vastly different, both were imprisoned for alleged subversion of those military projects.

Unlike Korolev, who suffered hard labor at the notoriously brutal Siberian camp, von Braun was incarcerated by his U.S. captors under incomparably more comfortable conditions in Fort Bliss, Texas, for a mere two weeks following Germany's WWII defeat.

Following his release along with 126 of his former Peenemunde colleagues, von Braun rose to a level of deserved international fame that Korolev would never know. Accomplishments of the American-German team he led included the development of the Jupiter intermediate-range ballistic missile, the Redstone rocket that launched America's first satellite and first U.S. astronaut Alan Shepard and the Saturn V rocket that enabled 12 fellow Earthlings to walk on the Moon.

Wernher von Braun was born of a noble family on March 23, 1912, in Wirsitz (now Wyrzysk), Poland, which was at that time part of Prussia and the German empire. His rocket interest was kindled by Transylvanian rocket pioneer Hermann Oberth's 1923 *"By Rocket into Planetary Space"* (English translation). A decade later, while pursuing a mechanical engineering degree, his university VfR "Spaceflight Society" conducted liquid-fueled rocket motor tests in support of Oberth's work.

Crediting Hungarian engineer Oberth as an important

career mentor, von Braun wrote:

> *Hermann Oberth was the first, who when thinking about the possibility of spaceships grabbed a slide-rule and presented mathematically analyzed concepts and designs...I, myself, owe to him not only the guiding-star of my life, but also my first contact with the theoretical and practical aspects of rocketry and space travel. A place of honor should be reserved in the history of science and technology for his groundbreaking contributions in the field of astronautics.*

Von Braun subsequently pursued a doctorate in physics at the University of Berlin, graduating in 1934. That same year his academic group launched two rockets reaching between one and two-mile altitudes. His graduate studies included rocketry research conducted at a solid-fuel rocket station not far from Berlin under the supervision of then-Captain Walter Dornberger, a department head for the German armed forces Ordinance Department.

During this time period, the National German Workers Party (NSDAP, or Nazi party) came into power and moved rocketry into the national agenda. His 1934 thesis titled *Construction, Theoretical, and Experimental Solution to the Problem of Liquid Propellant Rocket* was kept classified by the German government and not published until 1969.

In the early 1940s, von Braun moved to the new Peenemunde facility as its technical director under the command of Captain Walter Dornberger where his group, in combination with the Luftwaffe, developed liquid-fuel rocket engines for aircraft and jet-assisted takeoffs. Even more significantly, Peenemunde became the development center for a new A-4 ballistic missile which was to become better known as the V-2.

Von Braun was briefly imprisoned on espionage charges for resisting an attempt by Gestapo Chief Heinrich Himmler to take control of the V-2 project. He was reportedly released under Hitler's personal order soon after Germany invaded Poland to start World War II in 1939.[clxx]

A severe wartime labor shortage in 1943 prompted a plan to use slave labor at the Peenemunde V-2 rocket factory located at Mittlelwork. As with other slave labor operations, brutal treatment of working prisoners produced many tragic casualties. Although von Braun admitted visiting the plant on many occasions and called conditions there "repulsive," he claimed never to have witnessed any beatings or deaths directly. However, he admitted that by 1944 it had become clear to him that these incidents had, in fact, occurred.[clxxi]

Adolf Hitler signed an order on December 22, 1942, approving mass production of the V-2 to target London. British and Soviet intelligence agencies soon became aware of the program. Over the nights of August 17 and 18, 1943, the RAF Bomber Command's "Operation Hydra" dispatched 596 aircraft which dropped 1,800 tons of explosives on the Peenemunde facility. Although it was later salvaged and most of von Braun's team escaped unharmed, the raids killed his engine designer and chief engineer and succeeded in interrupting the program.[clxxii]

Historian Michael Neufeld quotes von Braun in his book *Wernher von Braun: Dreamer of Space, Engineer of War* expressing unhappiness upon hearing news of the London raids. Representing his interests in rocket applications for space travel rather than war, he reportedly said "the rocket worked perfectly, except for landing on the wrong planet." [clxxiii]

Following the end of WWII in 1945, von Braun and his

rocketry team (including his brother Magnus) voluntarily surrendered to American forces as part of "Operation Paperclip," and he eventually became technical director of the U.S. Army Ordnance Guided Missile Project in Huntsville, Alabama, as well as director of the NASA Marshall Space Flight Center from 1960 to 1970. He also later became vice president of the aviation company Fairchild Industries Inc. and a National Space Institute founder.

In addition to crediting inspirational influences of Hungarian Hermann Oberth, von Braun also acknowledged the importance of lessons taken from technical journals of an American rocketry pioneer named Robert Goddard. Commenting on Goddard designs, he observed:

> *His rockets…may have been rather crude by present-day standards, but they blazed the trail and incorporated many features used in our most modern rockets and space vehicles.*[clxxiv]

German and Russian liquid-fueled rocket development began years after Robert Goddard launched the world's first one weighing 16 pounds on March 16, 1926, from his Aunt Effie's farm in Auburn, Massachusetts. This was five years before Johannes Winkler launched Germany's first one at Dessau in 1931, and seven years before the Soviet Union's GIRD-09 1933 success in the Nakhahino woods.

As Goddard described his launch to sponsor Charles G. Abbot at the Smithsonian Institution:

> *After about 20 seconds the rocket rose without perceptible jar, with no smoke and with no apparent increase in the rather small flame, increased rapidly in speed, and after*

*describing a semicircle, landed 184 feet from
the starting point—the curved path being due
to the fact that the nozzle had burned through
unevenly, and one side was longer than the
other. The average speed, from the time of
flight measured by a stopwatch, was 60 miles
per hour. This test was very significant, as it
was the first time a rocket operated by liquid
propellants traveled under its own power.*[clxxv]

Although each of these early launches entailed extremely tiny
and crude devices by today's standards, paraphrasing the
immortal words of Neil Armstrong upon reaching the Moon's
surface on July 20, 1969, those small steps indeed led to giant
leaps.

Some financial support from the Guggenheim family—
thanks to an endorsement from American aviation hero Charles
Lindberg—enabled Robert Goddard to leave his professorship
position at Clark University in Worcester, Massachusetts, and
move his rocket development work to Roswell, New Mexico,
in 1931. However, lack of success in obtaining U.S. military
interest along with bad economic depression conditions which
reduced existing sponsorship support forced him to return to
academia. Ironically, only 12 days after the first successful
GIRD launch, Goddard received a letter from the Acting Navy
Secretary stating:

*Because of the great expense that would be
entailed in development of the rocket principle
for ordinance and aircraft propulsion, which
under present stringency of funds appears
hardly warranted, the Department regrets it is
not in a position to further such*

development.[clxxvi]

Goddard received a similarly discouraging rejection letter seven years later in 1940 from the U.S. Army Air Corps. A letter from Brigadier General H. Brett stated:

> *While the Air Corps is deeply interested in the research work being carried out by your organization under the auspices of the Guggenheim Foundation, it does not, at this time feel justified in obligating further funds for basic jet propulsion research and experimentation.*[clxxvii]

Goddard tended to eschew publicity, sharing many of his most imaginative ideas only with trusted friends and groups. He did, however, publish a March 1920 letter to the Smithsonian which discussed possibilities of photographing the Moon and planets from rocket-powered fly-by probes, sending messages to distant civilizations on inscribed metal plates, the use of solar energy in space, and the idea of high-velocity ion propulsion.

Those early ideas, which were generally regarded as very radical at the time, drew strongly sensationalized media publicity and criticism. A front-page January 12, 1920, *New York Times* story titled *Believes Rocket Can Reach Moon*, was followed days later with an editorial that scoffed at Goddard's proposals. The article argued, among other disagreements, that:

> *[A]fter the rocket quits our air and really starts on its longer journey, its flight would be neither accelerated nor maintained by the explosion of the charges it then might have left. To claim that it would be is to deny a*

> *fundamental law of dynamics, and only Dr.*
> *Einstein and his chosen dozen, so few and fit,*
> *are licensed to do that.*

Then, to add more insult to injury, the *New York Times* challenged Goddard's understanding of Newton's fundamental laws. Asserting that thrust can't occur in a vacuum it concluded:

> *That Professor Goddard, with his "chair" in*
> *Clark College and the countenancing of the*
> *Smithsonian Institution, does not know the*
> *relation of action and reaction, and of the need*
> *to have something better than a vacuum*
> *against which to react—to say that would be*
> *absurd. Of course he only seems to lack the*
> *knowledge ladled out daily in high*
> *schools.*[clxxviii]

Although Robert Goddard's rockets never achieved great altitudes, that wasn't really his goal. Rather, his work concentrated upon perfecting liquid-fueled engines along with reliable and accurate guidance and control subsystems which would eventually achieve high altitudes without tumbling in the thin atmosphere and provide stability for sensitive experiments and other payloads current and future rockets would carry.

The father of American spaceflight was on the verge of developing larger rockets capable of reaching extreme altitudes when World War II intervened to change everything. And yes, such devices later proved to work very well in the vacuum of space after all.

Technologies of Peace and War

Peacetime innovations have inevitably found offensive and

defensive military applications throughout human history since the Chinese invention of gunpowder fireworks. Although none have come to be as profoundly impactful in recent times as rocketry, which was proceeded by flight human, the quantum-computing-enhanced weaponization of artificial intelligence which will be discussed later in this story may soon gain equal or greater transformative influence.

Ironically, these same innovations and technologies of horrific havoc also continue to enrich and empower humanity with more abundant necessities and increasing conveniences, such as electrification, food and industrial production, air travel and global Internet...to name but a few.

Twentieth-century electrification powered industrial mass production of new military tanks, submarines, airplanes and other armaments of World Wars I and II.

Thanks to new military technologies and the horrors of trench warfare, World War I (1914-1918) saw unprecedented levels of carnage and destruction. Also referred to "the first modern war," it introduced the early development and mass deployment of numerous new types of weapons that continue to be in use today. Included are the machine gun, U-Boats and deadly gases by the Germans; the tank by the British; and aerial combat aircraft and armament developments by both sides.

German World War I veteran Erich Maria Remarque characterizes some horrific human consequences from his individual perspective in his 1929 novel, *All Quiet on the Western Front*:

> *A man cannot realize that above such shattered bodies there are still human faces in which life goes its daily round. And this is only one hospital, a single station; there are hundreds of thousands in Germany, hundreds of thousands*

> *in France, hundreds of thousands in Russia.
> How senseless is everything that can ever be
> written, done, or thought, when such things
> are possible? It must be all lies and of no
> account when the culture of a thousand years
> could not prevent this stream of blood being
> poured out, these torture chambers in their
> hundreds of thousands. A hospital alone shows
> what war is.*[clxxix]

Tragically, as for being "The War to End All Wars," this was not to be the case. A century of devastating 20th-century conflicts had only begun.

The harnessing and unleashing of enormous energy stored in atoms, in combination with aerial bombers, ushered in weapons of previously unimaginable horrors of death and destruction that played a major role in ending World War II.

Airplanes first gained true military importance in Italy for reconnaissance, bombing and artillery correction flights in Libya during their 1911-1912 war with Turkey. Bulgaria followed with bombing attacks on Ottoman positions during the First Balkan War of 1912-1913.

World War I witnessed major offensive, defensive and reconnaissance airplane uses both by Allies and Central Powers. Opposing pilots began shooting at one another, and in late 1914, Roland Garros of France came up with the deadly idea of attaching a fixed machine gun to the front of his plane. The first aerial factory was scored on July 1, 1915, by German pilot Lieutenant Kurt Wintgens flying a purpose-built fighter plane featuring a synchronized machine gun.

Air-to-air combat became the making of legendary heroics. German ace Manfred von Richthofen, better known as the Red Baron, shot down 80 planes. René Paul Fonck on the

Allied side was credited with 75 aerial victories.

Aircraft technology between World War I (1919) and World War II (1939) rapidly evolved from low-powered wood and fabric biplanes to sleek high-powered aluminum single-winged craft.

World War II not only rapidly increased the pace of aircraft development and production, but also that of more precise and lethal flight-based weaponry used in strategic large-scale bombing campaigns and dive bomber attacks on small targets. New technologies such as radar and communication systems for coordinated air defenses accompanied these accelerating developments.

In 1942, Germany introduced the first operational jet aircraft (Heinkel HE 178), and in 1943, also produced the first jet bomber (Arado Ar 234). Germany also developed the first cruise missile (V-1), the first ballistic missile (V-2) and the first operational rocket-powered combat aircraft (Me 163). However, late introduction, fuel shortages and a declining war industry limited overall German jet and rocket-powered aircraft advantages.

The immediate post-World War II era saw great advancements in jet and rocket-powered flight. American Chuck Yeager broke the sound barrier in 1947 in the rocket-powered Bell X-1. Jet aircraft broke distance barriers in 1948 and 1952, first crossing the Atlantic, and then flying non-stop to Australia.

The Korean War saw extensive air-to-air combat and bombing missions. U.S. fighters are estimated to have shot down as many as 700 Soviet air-combatants. Most of these dogfights took place over enemy-controlled areas.

The Vietnam War witnessed a strategic emphasis upon combat with air-to-air missiles. Close-proximity dogfights became less frequent.

The invention of nuclear bombs increased the strategic importance of military aircraft during the Cold War between the East and West. At first, supersonic interceptor aircraft were produced in great numbers by both sides to counteract devastating threats posed by even a small fleet of long-range bombers. By 1955, this emphasis shifted to surface-to-air missiles; then later again to prioritize intercontinental ballistic missiles capable of deploying nuclear warheads.

Any discussion of important human milestone aviation innovations must also include helicopter developments. Although the original general concept dates back to Leonardo da Vinci, reliable helicopters capable of stable hover flight were developed decades after fixed-wing aircraft. This circumstance is largely due to a requirement for more power versus weight requirements. Improvements in engines and fuels during the first half of the 20th century were a critical factor in making helicopters practical for modern warfare and civilian applications.

In 1885, Thomas Edison had attempted to build a helicopter powered by an internal combustion engine fueled by guncotton, an explosive. Explosions of the demonstration damaged the craft and badly burned one of his workers. Edison later patented a helicopter concept powered by a gasoline engine which never flew.[clxxx]

Frenchman Etienne Oehmichen set an early helicopter record in 1924 with a four-rotor craft which flew 1,180 feet. German engineer Heinrich Focke designed and built the first practical twin-rotor concept which in 1937 broke all previous helicopter records.[clxxxi]

Nazi Germany developed and used small numbers of helicopters during World War II for observation, transport and medical evacuation. Extensive bombing by Allied forces limited their production capacity.

In the United States, Russian-born engineer Igor Sikorsky developed the first practical lifting helicopter design. Produced for the military primarily for search and rescue during World War II, the craft had a single main rotor, along with a smaller rotor mounted on the tail boom to counteract torque produced by the larger one.

A key helicopter technology breakthrough occurred in 1951 when Charles Kaman applied a new kind of turboshaft piston engine developed in Germany to reduce weight and improve efficient performance. The lightweight turboshaft design led to the development of larger, faster and higher-performance helicopters, while many smaller and less expensive helicopters still use piston engines.[clxxxii]

"Medivac" for emergency medical airlift use was pioneered during the Korean War which dramatically reduced the previous average time needed to reach a medical facility during World War II and the Vietnam War. Military applications now also make extensive use of helicopters mounted with missile launchers and mini-guns to conduct aerial attacks on ground targets, as well as to ferry troops and supplies where the lack of an airstrip makes transport via fixed-wing aircraft impossible.

The development of rocketry, surveillance radar and electronic guidance systems established a new era of "push button" surface-to-surface, surface-to-air and air-to-air warfare that dominated proxy wars in Korea and Vietnam.

During the summer of 1940, Hitler launched massive Luftwaffe bombing attacks against the British Isles (the Battle of Britain). The terror directed against civilian targets included the V-1 flying bomb—also known to the Allies as the "buzz bomb"—the first of Germany's so-called "vengeance weapons" designed to demoralize London citizens.

While first suffering great devastation, Great Britain's

Royal Air Force eventually turned the air war against the aggressors, shooting down 2,698 German planes while losing only 915. This reversal marked the first of Hitler's major defeats.[clxxxiii]

In the winter of 1944, Hitler made a last desperate and failed war gamble in what is known as the Battle of the Bulge. The introduction of new, more modern Allied tanks and disadvantage of dwindling German troop numbers were decisive influences. Nevertheless, the casualties were terrible on both sides, making it one of the bloodiest battles of the war. It was also one of the costliest in all of American Army history.

Impressive advances of new military weapons had arrived too late to change the tide of war. Included was the replacement of the V-1 flying bomb with a faster V-2 flying bomb providing a larger payload jet aircraft which was vastly superior to propeller models, and submarine improvements which might have changed decisive outcomes of many Atlantic naval battles.

Following the Japanese pre-World War II bombing of Pearl Harbor, the first major Allied forces offensive against the Japanese Empire was the Battle of Guadalcanal launched primarily by U.S. Marines between August 1942 and February 1943.

Although by 1944 the Imperial Japanese Navy had lost nearly all of its defensive power, the Empire was determined to make American and Allied forces suffer more than they could endure. The Japanese fought to the last man, killing 6,800 Marines, and wounding nearly 20,000 more. Japanese losses were even greater, totaling well more than 20,000 men killed.

On August 6, 1945, the United States dropped an atomic bomb on the Japanese city of Hiroshima. Following the bombing, President Harry Truman issued a press release warning the Japanese either to surrender "…or expect a rain of ruin from the air, the like of which has never been seen on this Earth."

Truman didn't exaggerate. Three days later, the United States dropped a second atom bomb on Nagasaki. Between 140,000 and 240,000 citizens of the two cities perished.

America had suffered terrible losses as well. The estimated 426,000 human casualties included 161,000 dead (111,914 in battle and 49,000 non-battle) and 16,358 captured (not counting POWs who died). The United States also lost 21,355 aircraft, along with nearly 200 warships, including 5 battleships, 11 aircraft carriers, 25 cruisers, 84 destroyers and destroyer escorts and 63 submarines.[clxxxiv]

Supreme Commander of Allied forces during World War II, General Dwight D. Eisenhower, had recognized untenable social and economic war burdens on all of humanity:

> *Every gun that is made, every warship launched, every rocket fired signifies in the final sense, a theft from those who hunger and are not fed, those who are cold and are not clothed. This world in arms is not spending money alone. It is spending the sweat of its laborers, the genius of its scientists, the hopes of its children. This is not a way of life at all in any true sense. Under the clouds of war, it is humanity hanging on a cross of iron.*[clxxxv]

World War II had very substantially reshaped global power structures in another way: the introduction of incredibly powerful and horrifically devastating nuclear warfare capabilities. In its aftermath, evolutionary rocketry evolution lead to intercontinental ballistic missiles capable of delivering nuclear and thermonuclear devices to all points on the planet.

Failure of the League of Nations to prevent World War II led to its dissolution. It was replaced by a new United Nations organization on October 24, 1945, in a renewed attempt to maintain world peace. By 1946, there were 35 UN member states. Joined by newly independent nations, that membership grew in number to 127 by 1970.

The 1945 Yalta Conference agreement divided Western capitalist powers and the communist Soviet Union into separate European spheres of influence which set the stage for a geopolitical rivalry that would come to dominate international relations.

A Soviet iron curtain soon descended across the continent from Stettin in the Baltic, to Trieste in the Adriatic. Behind that line were located all capitals of the ancient states of Central and Eastern Europe: Warsaw, Berlin, Prague, Vienna, Budapest, Belgrade, Bucharest and Sofia.

Over the course of the war, the Soviet Union had already annexed several countries as Soviet Socialist Republics (SSRs). Eastern Poland was incorporated into Belarusian and Ukrainian SSRs; Latvia, Estonia and Lithuania became SSRs; part of eastern Finland became a Karelo-Finnish SSR; and eastern Romania became a Moldavian SSR.

Then, between 1945 and 1949, Yugoslavia, Albania, Bulgaria, Poland, Romania, Czechoslovakia, Hungary and East Germany became independent communist People's Republics with close Soviet ties as de facto satellite states.[clxxxvi][clxxxvii]

The rise of communism also spread outside Europe, adding the nations of Mongolia, China, North Korea and Vietnam into its fold. This expansion of communist ideology and Soviet influence created a deep and lasting rift between many former World War II allies. The two emerging rival blocks coalesced into formal competing mutual defense organizations, forming the North Atlantic Treaty Organization

(NATO) in 1949 and the Warsaw Pact among the USSR and its seven satellite states of Central and Eastern Europe in 1955.[clxxxviii]

Meanwhile, with post-war Western relations rapidly deteriorating, the Soviet Union, supported by espionage efforts, developed and detonated its first nuclear weapon in August 1949. The United States countered through a crash program to create the first hydrogen bomb in 1950 and detonated an even more destructive second-generation thermonuclear weapon in 1953 which was more than 400 times as powerful as those dropped on Japan. The Soviet Union then followed suit, detonating a primitive thermonuclear weapon in 1953, and a full-fledged version in 1955.

Development of computerized long-range nuclear delivery systems by both camps produced a rapidly accelerating and increasingly dangerous "mutually-assured-destruction" (MAD) Soviet versus United States arms race. Tensions led to a broader proliferation of nuclear weapon development and stockpiling. Several other nations, including the United Kingdom, France, China, India, Pakistan, North Korea and Israel are believed to have gained first-strike and retaliatory capabilities.[clxxxix]

Major post World War II communist-capitalist territorial power reshuffles and conflicts in combination with terrifying nuclear weaponry and long-range rocket delivery systems led the Cold War world ever-closer to the brink of MAD that such developments were at least theoretically intended to prevent. Holocaust was narrowly averted in the aftermath of a tense 13-day October 1962 political and military standoff between U.S. President John Kennedy and Soviet leader Nikita Khrushchev over missiles being shipped to Cuba.

The "Cuban Missile Crisis" occurred after Lockheed U-2 spy planes revealed missile launchers being installed over the

U.S. neighbor island which was controlled under Fidel Castro's socialist government with close Soviet Union ties. In response, Kennedy instituted a naval blockade around Cuba to block Soviet missile shipments. Threatening to penetrate the defense, military conflict was avoided when the USSR backed down and agreed to remove the missiles in exchange for a U.S. commitment not to invade Cuba.

The East-West Cold War that ensued was to lead to the most revolutionary pioneering adventure in human history.

PIONEERING A VAST SPACE FRONTIER

CONCEIVED IN THE genius of Konstantin Tsiokovsky, Hermann Oberth, Robert Goddard and other visionaries, mankind gave birth to a pioneering dream that would come to achieve inspirational goals previously unimaginable but a few decades ago. Orbiting satellites erased communication boundaries world-wide, spawned a transformative Internet information-sharing network, monitor natural and man-made events that affect our safety, coordinate and guide air and surface transportation movements, and support unlimited business opportunities.

Advancements in rocketry, spacecraft and instruments of exploration have opened an epic era of cosmic discovery. And yes, the complex challenges driving such achievements have yielded countless technological advancements that continue to enhance the quality of our everyday lives and expand our human experience.

Such developments followed unseen results arising from a far different motivation, one prompted by events originating in the former USSR that harshly jolted the psyche of America and the West.

An October 4, 1957, front-page *New York Times* headline in half-inch capital letters carried a story that was being reported all over the world: *SOVIET FIRES EARTH SATELLITE INTO SPACE; IT IS CIRCLING THE GLOBE AT 18,000 MPH; SPHERE TRACKED IN 4 CROSSINGS OVER US.* That orbit repeated more than 1,400 rounds before Sputnik-1 stopped chirping out its ominous presence and burned up in the atmosphere three months later.

Or as an October 7th Manchester Guardian editorial titled *Next Stop Mars* described the event:

> *The achievement is immense. It demands a psychological adjustment on our past towards Soviet society, Soviet military capabilities and—perhaps most of all—to the relationship of the world with what is beyond.*

It went on to more ominously speculate that:

> *The Russians can now build ballistic missiles capable of hitting any chosen target anywhere in the world.*

The concept of launching satellites to orbit wasn't new. Tsiolkovsky's 1903 calculations showed that a device launched at a certain velocity could overcome the pull of Earth's gravity and achieve orbit. Slightly more than a half-century later, Korolev's team had developed a rocket capable of accessing that necessary 8,000 meters per second orbital trajectory.

That goal had been studied and pursued in America as well. During the mid-1940s, the U.S. Army Air Corps asked major airframe companies to submit secret competitive proposals for the design of an "Earth-orbiting satellite." This led to funding a newly formed "Project RAND" (Research and Development) in Santa Monica, California, to study the matter. Their report concluded that:

> *The achievement of a satellite craft by the United States would inflame the imagination of mankind, and would probably produce repercussions in the world comparable to the explosion of the atomic bomb.*

During early 1954, the United States began considering plans to place a small satellite in orbit as its 1957-58 International Geophysical Year (IGY) contribution. In response, Wernher von Braun's team at the Army's Redstone Arsenal in Huntsville began meeting with George Hoover of the Office of Naval Research to accomplish this goal using existing Army Ordnance weapons technology. Their proposed solution, "Project Orbiter," was to be an Army-Navy-Air Force design.

The Soviets were concerned about America launching an IGY satellite before they did. Towards the end of 1953, the R-7 rocket made by the Chief Designer's team could launch a 5-ton ICBM warhead and could also easily orbit a 1.5-ton satellite.

But what sort of satellite? The Soviet Academy of Scientists was presented with various options for IGY. One possibility was a living organism such as a dog. Another was to fly around the Moon and photograph the side hidden from Earth. The big priority, however, was to beat the Americans. Too ambitious of a plan would fail that purpose.

They finally settled with that plain polished metal sphere carrying only a radio transmitter, batteries and temperature-measuring instruments. It worked, and as a Pravda headline proclaimed, "World's First Artificial Satellite of Earth Created in the Soviet Union."

One month later on December 6th, the first American Vanguard Program launch attempt ironically designated "TV-3" (for Test Vehicle 3) failed before world television cameras. After rising but a few feet off the ground it ignominiously sagged back, buckled, burst into flame and tossed its tiny three-pound satellite still transmitting a short distance away. *Pravda* reproduced a front-page London *Daily Herald* photo showing the explosion with a superimposed headline which in translation read "OH, WHAT A FLOPNIK!"

Far more fortunately for the United States, following a Sputnik-2 launch, a von Braun group from the Army Ballistic Missile Agency developed a "Jupiter C" launch vehicle which successfully placed a 28-pound "Explorer-1" satellite in orbit on January 31, 1958. The scientific benefits proved historic when its onboard instruments first discovered now famous Van Allen radiation belts.

Technological bragging tables turned on the Russians after a launch failure three months later on April 27, 1958, with its 1.3-ton Sputnik-3 payload aboard. Although a successful follow-up launch was soon achieved, its replacement Sputnik-3 satellite missed an opportunity to further map the Van Allen belts due to a satellite positioning failure.

Following Sputnik, a seesaw series of competitive orbital launch successes and flubs ensued on both sides.

The first three Soviet attempts to place satellites on the Moon failed to reach an Earth-departure orbit. The fourth, a January 2, 1959, "Luna-1" launch missed its target by 6,000 kilometers yet succeeded in orbiting the Sun. Luna-2 launched on September 12, 1959, and became the first spacecraft to make contact with the Moon or any other celestial body. Luna-3, launched only three weeks after Luna-2, photographed the far side of the Moon never before seen by humans.

Meanwhile, Americans were realizing some launch misses and hits as well. These began on August 17, 1958, when a first stage Air Force Thor-Able spacecraft of the subsequent Pioneer Program malfunctioned 77 seconds after launch from Cape Canaveral. Although Pioneer-1 third stage missed the Moon, it set a distance record by traveling some 113,854 kilometers into space. Pioneer-3 then provided important data about the outer Van Allen radiation belt. Notwithstanding these significant achievements, there were seven straight Pioneer Moon misses through 1960.

As for moving targets, Mars and Venus destinations presented far more complicated trajectory guidance challenges than the Moon. Two October 1960 Soviet Mars probe failures to reach Earth's orbit were followed by seven straight failed Venus probes—five by Russia and two by America—between February 1961 and September 1962.

The first failed Soviet Mars launch. which occurred at the time of the Khrushchev-Kennedy standoff over Cuban missile emplacements, might very well have led to a little-publicized but hugely larger competitive Russian-United States disaster. Just as the spacecraft was being prepared for launch, Korolev's team was ordered to immediately remove it and abort the mission so that a military ICBM could use the site in response to a U.S. thermonuclear strike.

Although the issue soon appeared to be settled "via diplomatic channels" and launch preparations were allowed to proceed, that didn't end the problem.

On October 24, 1962, still in the middle of the crisis, the Mars launcher exploded into so many pieces during ascent that observers at the U.S. Ballistic Early Warning System feared that a Soviet nuclear attack might have commenced. The crisis was averted when computers which assess trajectory and impact points reported a false alarm within seconds.

The international nuclear war scares subsided on October 27, 1962, when Khrushchev announced he would dismantle the missiles in Cuba and return them to the USSR. One week later a Soviet spacecraft designated Mars-1 made the first (unintentional) Red Planet flyby after losing communications. Nevertheless, it accomplished the impressive feat of traveling 106 million kilometers and sending back 61 batches of data until March 21, 1963.

The mutually disclaimed U.S.-Russian competition intensified over the next several years, with each nation anticipating and closely monitoring activities of the other. Both continued to experience significant, if incremental, successes and failures.

A Russian Zond-1 spacecraft launched on April 2, 1964, and reached the vicinity of Venus on July 20, although a radio failure resulted in no returned data. Venera-3 launched on November 16, 1965, and accomplished the first Venus impact on March 1, 1966. Zond-2 launched on November 30, 1964, and demonstrated use of the first electric thrusters for attitude control.

On the American side, although its instruments failed, on April 23, 1962, Ranger-4 became the first U.S. spacecraft to impact the Moon. Launched on October 18 of that year, Ranger-5 missed the Moon by 700 kilometers. Ranger-6 hit the Moon on January 30, 1964, but the TV camera didn't work.

Rangers 7, 8 and 9 returned marvelous pictures covering over 400,000 square kilometers of the lunar surface between 1964 and 1965. And while Mariner-1 failed to reach Venus, Mariner 2 flew within 35,000 kilometers of the planet. Mars-3 missed Mars, but in July 1965, Mars-4 sent back spectacular TV pictures of its cratered surface from a distance of 9,844 kilometers.

The 1970s witnessed more historic interplanetary achievements by both Russia and the United States. In 1971, five years after Korolev's death, two Soviet capsules released by Mars-2 and Mars-3 crashed into the Martian surface on November 27 and December 2, respectively. The first of those events occurred just two weeks after the American Mariner-9 developed by NASA's CalTech Jet Propulsion Laboratory orbited around the planet throughout a dust storm, then sent back detailed pictures of the surface until January 1972.

A New Human Exploration Domain

Although Korolev never lived to witness Russian probes reaching Mars, he did experience a personal triumph which commenced a transformational new era of human space exploration. As the *New York Times* exclaimed on April 12, 1961, once again in bold front-page headlines:

> *SOVIET ORBITS MAN AND RECOVERS HIM; SPACE PIONEER REPORTS: 'I FEEL WELL': SENT MESSAGES WHILE CIRCLING EARTH.*

The newspaper followed up with an editorial the next day prophesying that the "flight will be hailed as one of the great advances in the story of man's age-old quest to tame the forces of nature." Pravda declared it a "GREAT EVENT IN THE HISTORY OF MANKIND." The Communist Party seized upon the event as a triumph over capitalism.

This was not generally greeted as good news by the majority of Americans, and particularly not by those connected with the U.S. space program. Gagarin's one-hour, 48-minute full-orbit demonstration aboard a Vostok-2 spacecraft on April 12, 1961, eclipsed a 15-minute-long suborbital launch of Navy jet pilot Lt. Col. Alan Shepard on May 5th of that year which reached a 167-mile altitude and traveled 302 miles downrange.

Unfortunately for American history, Shepard's flight, which was originally intended to be a full-orbit launch aboard a Redstone rocket, was delayed multiple times following von Braun decision that another test flight was needed after a previous one traumatized its passenger...a chimpanzee named Ham.

George Low, then chief of manned space flight, recalled a conversation between newly appointed NASA Administrator James Webb and his deputy Robert Seamans, who had just testified on the state of their efforts before the House Committee on Science and Astronautics on the day before Gagarin's flight. Webb and Seamans decided not to show a film following Gagarin's historic world spectacle. The movie featured recovery of the dazed chimpanzee Ham two and one-half months earlier.

Low remembered the conversation prudently concluding:

> ...it would not be in our best interest to show
> how we had flown a monkey on a suborbital
> flight when the Soviets had orbited Gagarin.

Although the distinction of being the first American to orbit Earth ultimately went to John Glenn, fortune later beamed more brightly upon Alan Bartlett Shepard Jr. upon becoming the only Mercury astronaut to walk on the Moon on the Apollo 14 mission.

Gus Grissom suffered far worse fortunes. He nearly drowned when explosive bolts fired unexpectedly, blowing the hatch off his "Liberty Bell" capsule during splashdown following the second suborbital Project Mercury-Redstone flight on July 21, 1961. A catastrophic fire ended the lives of Apollo

Astronauts Grissom, Roger Chaffee and Ed White during a January 27, 1967, test.

Russia and America had both been conducting animal tests to determine if humans could survive launch and re-entry stresses in addition to weightless orbital conditions. Soviet space scientists had been experimenting with canines since at least 1951. Dogs Dezik and Tsygan were sent to a 100-kilometer altitude that year using the same pod that carried Laika. The United States had experimented with monkeys since the 1950s.

Not known to Westerners until mentioned in a 1994 publication, Russian space canine experiments had not always led to successes. A July 28, 1960, Vostok prototype flight carrying dogs Chaika and Lisichka failed when the launch vehicle exploded. More fortunately, an 18-orbit flight the following month ended far better for dogs Belka and Strelka, who became the first creatures to return alive.

A few weeks later, the Communist Party had approved a request to launch a human. Not even the catastrophic October 24, 1960, launch pad explosion nor the failure of another December 1st launch failure which killed two other dogs halted plans to go ahead.

As always, top-secret Chief Designer Korolev was kept off to the side away from public fanfare following Gagarin's triumphant return. Worse for America, NASA achievements had been sidelined altogether. A chastened America got the Gagarin flight message without need of the dazed monkey film.

Revelations of Soviet ballistic missile advancements and geopolitical implications had made international headlines in 1962 when intended placements in Cuba of R-16 ICBMs which triggered a fearsome Kennedy-Khrushchev confrontation. Their range of about 2,200 kilometers combined with their basing in Cuba and western Russia posed a major threat not only to the

United States, but also to bomber bases in Europe and Asia.

Both countries at that time were aware they were being spied upon from overhead by the other. Korolev's design bureau had developed and launched four spy satellites beginning with Kosmos-4 (later known as "Zenit") on April 26, 1962. By that time, the United States had already been flying its own spy satellites for nearly two years under a top-secret "Corona" program that was declassified in 1995.[cxc]

The Zenit cameras could cover the entire United States in about twenty-five orbits and reportedly determine the number of cars in a parking lot. These were ideal capabilities for ICBM site mapping and monitoring. Corona could presumably do the same.[cxci],[cxcii]

America Races to the Moon

The combined timing of the Bay of Pigs and Gagarin headlines put great pressure on Kennedy to demonstrate resolute leadership. On May 25, 1961, just slightly less than six weeks following Gagarin's catapult into the Space Hall of Fame, he did so, announcing before a special joint session of Congress that an American astronaut would be safely sent to the Moon "before this decade is out."

The American national security implications of both events leading up to Kennedy's bold declaration were made crystal clear in Soviet leader Nikita Khrushchev's November 22, 1957, statement during an interview with publisher Randolph Hearst Jr. three months after his country had conducted a R-7 missile simulated warhead launch:

> *The Soviet Union possesses intercontinental ballistic missiles. It has missiles of different systems for different purposes. All our missiles can be fitted with atomic and hydrogen*

warheads. Thus, we have proved our
superiority in this area.[cxciii]

Even more pointedly, Khrushchev disparaged the defensive
potency of U.S. naval power in a September 7, 1958, letter to
President Eisenhower, stating:

> *In the age of nuclear and rocket weapons of*
> *unprecedented power and rapid action, these*
> *once formidable warships are fit, in fact, for*
> *nothing but courtesy visits and gun salutes, and*
> *can serve as targets for the right type of*
> *rockets.*[cxciv]

The issue of a purported U.S.-Soviet "missile gap" had entered
the 1960 presidential campaign when candidate Kennedy
claimed that the Russians had built up a substantial lead.
President Eisenhower, who had access to spy satellite and
secret U-2 overflights, knew differently, but refrained from
saying so to protect those sources. Early 1960 CIA Russian
ICBM estimates put the total at 35 missiles, with the number
growing from between 140 and 200 by mid-1961...not the
thousands that some were wildly speculating.[cxcv]

Conditions for American dominance in a missile race had
grown grim. In response to a reporter's question regarding
when the United States might *"*perhaps surpass Russia in this
field,*"* President Kennedy providently observed, *"the news will*
be worse before it is better." He was right. On August 6, 1961,
Gherman Titov's seventeen-orbit Vostok-2 flight topped Gus
Grissom's suborbital Redstone-4 flight of July 21.

Kennedy's September 12, 1962, commitment to human
lunar exploration at Rice University's stadium left no doubt
that this was to be a competitive race dedicated to

demonstrating U.S. technological supremacy over the Russians. Kennedy warned:

> *Within these last 19 months at least 45 satellites have circled the Earth. Some 40 of them were made in the United States of America and they were far more sophisticated and supplied far more knowledge to the people of the world than those of the Soviet Union.*

What had previously been a tacit matching of wits and capabilities involving post-war rivals had become an officially recognized race between superpowers. The situation literally began to look up for America when a Mercury-Atlas 6 rocket launch of Friendship-7 carried Colonel John Glenn on three orbits on February 20, 1962.

Three months later, Mercury-Atlas 7 carried Scott Carpenter on three orbits, splashing down 420 kilometers beyond the target area due to reentry errors. On October 3, 1962, Mercury-Atlas-8 Astronaut Walter Schirra did even better, splashing down within 7.24 km of a recovery ship following 6 orbits. That record, in turn, was beaten by Gordon Cooper on May 16, 1963, on Mercury-Atlas 9, which landed 6.4 km from ship following 22 orbits.

Meanwhile, on the other side, Soviet Cosmonaut Andrian Nikolayev had performed 65 orbits on August 11-15, 1962, aboard Vostok-3, while Pavel Popovich, who launched the next day aboard Vostok-4, did 48, the two spacecrafts flying in orbits within 5 kilometers of each other. On June 14-19, Valery Bykovsky, aboard Vostok-5, passed within 5 kilometers of Russia's (and the world's) first woman, Cosmonaut Valentina Tereshkova, aboard Vostok-6.

The high international prestige stakes of a race to the

Larry Bell

Moon competition weren't lost on Khrushchev, Korolev and others in the Soviet Union. As quoted by Korolev associate Oleg Ivanovsky:

> He would tell us that 'the Americans are at our heels, and the Americans are serious people.' He wouldn't use the word 'Amerikantsi' but 'Amerikan-ye' as if these weren't just American residents but the entire American culture we were competing with. He didn't mean this as an insult but as a show of respect for the competition.[cxcvi]

Korolev is believed to have made his first serious proposal for a manned Moon mission during an April 6, 1956, speech to the Soviet Academy of Scientists...more than a year prior to the Sputnik-1 launch. He stated:

> This real task is to fly to the Moon and back from the Moon. This task is most easily solved by starting from Earth. Somewhat more difficult will be returning to Earth that will be on a satellite or rocket that goes to the Moon. But it must not be believed that the proposals I am making are extremely remote.[cxcvii]

Korolev's writings made his primary objective clear:

> [These] first studies of the Moon and interplanetary space at distances that reach 400-500 thousand kilometers will also create the necessary prerequisites/premises for the penetration of man into interplanetary space,

198

the Moon and the planets.[cxcviii]

Buzz Aldrin told me prior to the release of news reports following Apollo that he believed that the Russians had secretly planned to beat America to the Moon with their astronauts. In 1988, I was among the first Americans to be invited, along with several of my space architecture graduate students, to visit the facility in Moscow and to see a mockup of a lunar surface habitat module they had been planning for exactly this purpose.

The Lunokhod ("Moonwalker") program, which was intended from the beginning to support manned surface missions, had produced remote-controlled rovers that predated all others to be deployed on a celestial body. Launched aboard powerful Proton-K rockets, they were controlled from a network of ten ground-based facilities containing Earth satellite vehicle tracking equipment along with command/controls for Soviet near-space civil and military operations.

The first of the total two successful Lunokhod rover deployments (Luna 17) reached the Moon in the Sea of Rains on November 17, 1970. Measuring 4 ft. 5 inches high, it carried four television cameras, special extendable devices to collect lunar soil for density and mechanical property tests and a cosmic ray detector. Over its 322 Earth days of operations, it traveled 6.5 miles, returned more than 20,000 TV images and performed a series of 25 X-ray fluorescence soil analyses at 500 different locations.

Lunokhod-2 (Luna 21), a more advanced robot, landed on January 15, 1973. Equipped with three slow-scan TV cameras mounted high on its rover for navigation, it returned images to ground controllers on Earth who sent real-time driving commands. Scientific instruments included a soil mechanics tester, solar x-ray equipment and a French-supplied photodetector for laser detection experiments. It returned

Larry Bell

about 80,000 pictures over five months of operations.

Major achievements of Russia's Moon program also included three robotic sample return missions: Luna-16 (September 1970), Luna-20 (February 1972) and Luna-24 (August 1976). Altogether they collected and returned slightly less than a pound of lunar surface materials.

Peaceful Pathways Forward

Although not broadly known, two years before his assassination on November 22, 1963, Kennedy had proposed a possible joint cosmonaut-astronaut lunar mission to Khrushchev at a Vienna luncheon. After some consideration, Khrushchev rejected the offer.

Kennedy proposed the idea again just two months before his death at a September 20, 1963, United Nations General Assembly appearance, stating:

> *Space offers no problems of sovereignty…Why, therefore, should man's first flight to the Moon be a matter of international competition? Why should the United States and the Soviet Union, in preparing for such expeditions, become involved in immense duplications of research, construction, and expenditure? Surely we should explore whether scientists and astronauts of our two countries—indeed of all the world—cannot work together in the conquest of space, sending some day in this decade to the Moon not the representatives of a single nation, but the representatives of all of our countries.*[cxcix]

In any case, while neither Kennedy nor Khrushchev would live to see their astronauts and cosmonauts joining together on evening Moon strolls, both nations would leave separate tracks. Twelve Apollo explorers beginning with Neil Armstrong and Buzz Aldrin would impress human footprints on its surface, while robotic Soviet explorers would leave mechanical track prints before the end of the Apollo program in 1974.

International pathways converged as the 1970s and 80s ushered in a new era of international space cooperation. With Apollo ended, priority Russian and U.S. attention shifted from lunar exploration to Earth orbital studies of human adaptation and mitigations associated. These studies focused on extended weightlessness, influences of weightlessness and space vacuum upon materials and physical/mechanical processes and ways to enhance human safety and performance during future long-term, multi-year missions.

Many nations accepted invitations to join and invest in common, peaceful enterprises of discovery in orbit and beyond. As a result, the world continues to witness benefits of multicultural engagement rising far above limited boundaries of interest defined on surface maps. This ongoing quest of discovery may ultimately prove to be the greatest and most rewarding space exploration challenge of all.

Global audiences watched on July 17, 1975, as TV images showed Soviet Cosmonauts Alexei Leonov and Valery Kubasov shaking hands with NASA Astronauts Thomas Stafford, Vance Brand and Donald ("Deke") Slayton high above the Atlantic Ocean. The historic docking of an American Apollo Command and Service Module (CSM) and Russian Soyuz vehicle mated critical mechanical pressure seals developed separately for the first time. This feat occurred after the two spacecrafts delivering them had made gradual trajectory changes over a two-day period after being launched within hours of each other. One had departed from the Kennedy Space Center in

Florida...the other from the Baikonur Cosmodrome in Kazakhstan.

The political timing of the Apollo-Soyuz mission was no accident, highlighting a new policy of détente, a symbolic act of peace, between superpowers. The United States was engaged in a Vietnam ground war which, given Russia's proxy involvement in the conflict, was adding to existing Cold War tensions. The government-controlled Soviet press had been highly critical of America's Apollo program, printing on one occasion that:

> ...the armed intrusion of the United States and Saigon puppets into Laos is a shameless trampling underfoot of international law.

Soviet leader Leonid Brezhnev shifted that public position to extol peaceful diplomatic benefits of the Apollo-Soyuz experiment. He told the world:

> The Soviet and American spacemen will go up into space for the first major joint scientific experiment in the history of mankind. They know that from outer space our planet looks even more beautiful. It is big enough for us to live peacefully on it, but is too small to be threatened by nuclear war.

Apollo-Soyuz, the first joint U.S.-Russian space flight, was to be the last flight for an Apollo spacecraft. It was also to be the last manned U.S. space mission until the first Space Shuttle launch in April 1981, providing important experience for future Shuttle-Mir and International Space Station (ISS) programs that followed.

What Makes Humans Truly Exceptional?

NASA's Space Shuttle program emerged when the agency convened a task group in 1968 to begin planning beyond Apollo. The priority centered upon developing a reusable Earth-to-orbit transportation system which could support a space station with round-trip crew and cargo delivery.

The program was formally launched by President Nixon's administration on January 5, 1972, with goals of "transforming the space frontier...into familiar territory, easily accessible for human endeavor." North American Rockwell (later Rockwell International, now Boeing), the same company responsible for building the Apollo Command/Service Module, was awarded the development contract.

Over the course of its operations between 1981 and 2011, the five-Shuttle-fleet program accomplished 135 missions. The longest, STS-80, lasted 17 days, 15 hours, and the final flight, STS-135, occurred on July 8, 2011.

Two of the Shuttle orbiter missions suffered catastrophic disasters which killed a total of 14 crew members. The STS-51-L Challenger launch, which exploded 73 seconds after liftoff on January 28, 1986, was caused by the failure of a connecting seal on a long solid-fuel rocket. STS-107 Columbia was lost approximately 16 minutes before its expected landing on February 1, 2003, when a piece of hard insulating foam shed from its large external tank during launch punctured a reentry heat protection tile on its wing edge.

The Soyuz modules originally developed to carry cosmonauts to Salyut space stations and Mir later became used for ISS. A minimum of two are docked with the ISS at all times to provide assured contingency departure for a crew of six (three passengers each).

The ISS core module providing its primary crew life support systems is based upon the engineering developed for the world's first space station, Salyut-1, which was launched in

1971. Salyut 1, in turn, was originally developed by Korolev's design bureau undercover for a military space station secretly known as "Almaz."

Cosmonauts from the Communist bloc and non-American astronauts from the West later repeatedly set time-in-orbit endurance records aboard Salyut 4, 6 and 7 from April 1982 up through early 1991.

Skylab, America's first space station, orbited Earth from 1973 to 1979 and was created using a converted third stage of a Saturn V Moon rocket outfitted with two decks as a habitat and orbital workshop. Strictly a NASA operation, the facility was spacious even by current standards and provided a large solar observatory and experiment area. A Crew Service Module (CSM) converted from the second stage, a smaller Saturn 1B booster, provided crew transport and emergency means to rapidly return to Earth.

Skylab's three total three-crew missions logged 513 man-days in orbit and accomplished thousands of experiments covering many different disciplines. Its orbit was allowed to slowly decay causing the facility to burn up on reentry five years after the last crew had returned home.

The Mir space station (which translated, combines words like "world," "peace" and "village") ultimately served as a true symbol of cooperation between the people of Russia and the United States following a half-century of mutual antagonism. Over its 15 years of operation from the time its first module was launched on February 20, 1986, Mir hosted 125 cosmonauts and astronauts from 12 different nations who conducted approximately 23,000 scientific experiments.

America's Space Shuttles docked with Mir seven times. While prior to the Shuttle-Mir missions, experiments and supplies were provided exclusively by Russian "Progress" cargo vehicles, several Shuttle-Mir missions supported commercial services.

As the world's first modular space station, Mir's cluttered outside appearance has been variously characterized as a dragonfly with wings outstretched, a prickly hedgehog and a 100-ton Tinker Toy. Nevertheless, Mir served as a beautiful symbol of international cooperation in space science and collegiality over a course of 86,000 total orbits.

Ironically, on the very same day Mir's life ended as fragments in a watery grave on March 23, 2001, Russia expelled four U.S. diplomats and threatened to expel 46 more in retaliation for America's expulsion of 50 of theirs for suspected espionage.

Mir's demise un-coincidentally coincided with the beginning of the most complex international scientific and engineering project in history and the largest human structure ever to be put in space.

Planned and operated by five different agency partners representing 15 different countries, the International Space Station serves a variety of purposes: a laboratory for biological, material and other sciences; an observation platform for astronomical, environmental and geological research; and a stepping-stone towards future space exploration. Responsibilities and investments are divided among NASA, Russia's Federal Space Agency Roscosmos, the European Space Agency, the Canadian Space Agency and the Japan Aerospace Exploration Agency.

The scale of ISS is immense, spanning the area of a U.S. football field including end zones. Weighing a total of nearly a million pounds and orbiting at five miles per second the complex now provides more livable space than a conventional five-bedroom house where a six-person expedition crew typically remains onboard from between four to six months. For comparison, ISS is nearly four times larger than Mir, and about five times larger than Skylab.

During a visitation with a docked Space Shuttle, the combined ISS complex has supported a total of 13 people for several days. This crew size was temporarily reduced to two-person teams after the tragic Columbia Shuttle disaster during which time the crew and supplies could only reach ISS using Russian Soyuz and Progress spacecraft. ISS hosted its first one-year crew in 2015-2016 involving NASA's Scott Kelly and Roscosmos Russia's Mikhail Kornienko.

International partner dependence on Russian Soyuz vehicles to transport its astronauts to and from the ISS following the end of the U.S. Shuttle program will end as new commercial crew launch and landing services are coming online. By 2012 the ISS was already being supported by commercial cargo delivery services provided by SpaceX's reusable Dragon spacecraft, followed by Orbital Science's Cygnus spacecraft in late 2013. Subject to safety certification, each company will be able to carry up to 7 passengers.

Founded in 2002 by PayPal co-founder Elon Musk, SpaceX (the "X" referring to Exploration Technology), has already logged several historic achievements: In 2008, SpaceX became the first privately funded company to launch a (Falcon 1) rocket into orbit; the first to successfully orbit and recover a spacecraft (2010); the first to send a spacecraft to the ISS (2012); the first to launch a satellite into LEO (2013); and the first organization, private or government, to successfully return a first stage back to the launch site and accomplish a vertical landing with a rocket on an orbital trajectory (2015).

SpaceX also conducted its first satellite delivery to geosynchronous orbit (NASA's Deep Space Climate Observatory) in 2015. In June 2018, a two-stage, 180-feet-tall SpaceX Falcon 9 booster conducted the first commercial cargo delivery flight to the ISS, and in March 2019, the company successfully launched and docked an unmanned Crew Dragon capsule to the ISS.

SpaceX is developing a larger reusable replacement of its Falcon 9 named the "Big Falcon Rocket" (BFR) with a lift capacity to Earth orbit of more than 150 tons. Also referred to by the company as an "Interplanetary Transport System" (ITS), its first test flight has been tentatively scheduled for 2020.

Founded in 1982, Orbital Sciences (Orbital ATK), which developed the Antares Rocket and Cygnus Spacecraft, helped to create the Orion Launch Abort System that would have helped NASA astronauts escape in the event of an emergency involving the now-canceled Ares 1 rocket. Orbital ATK is currently competing with SpaceX for ISS cargo resupply services using expendable foreign-supplied rockets and delivery capsules.

SpaceX was not the first commercial company to develop and demonstrate a reusable rocket booster rather than sacrifice it after a single flight. On January 23, 2016, Blue Origin, a company created by Amazon founder/CEO Jeff Bezos in 2004, accomplished this feat. The re-used launcher had been recovered from an earlier successful launch and landing that occurred only two months earlier on November 24, 2015, at Blue Origin's West Texas test facility.

Blue Origin's vertical take-off and landing rocket design is powered by the company's own BE-3 engine. The company has developed a sub-orbital six-person "New Shepard" tourist capsule named in honor of Alan Shepard, who piloted America's first suborbital mission. In addition, Blue Origin is also in the process of creating a two-stage heavy-lift orbital vehicle named the "New Glen" in honor of John Glenn. The first stage will be powered by seven of Blue Origin's new BE-4 engines.

NASA is also commissioning a Space Launch System (SLS) and an Orion crew spacecraft for missions beyond Earth Orbit. The SLS is described as a "super-heavy" expendable launch vehicle which is part of NASA's deep space exploration plans, which will eventually include a crewed mission to Mars. It is being built by Boeing, the United Launch Alliance, Northrop Grumman and Aerojet Rocketdyne.

The Orion Multi-Purpose Crew Vehicle (Orion MPCV) is a joint NASA-European Space Agency venture designed to carry a crew of four astronauts to destinations beyond low-Earth orbit in connection with such missions as to the Moon, asteroids and Mars. The overall vehicle consists two main modules: a command module which is primarily being built by Lockheed Martin and a service module containing many of the mechanical and life support systems which are being developed by Airbus Defense and Space.

Several other nations, including some which are relatively new to space era activities, are also embarking upon ambitious plans to explore this new frontier of scientific, technological and economic opportunities both in Earth's orbit and beyond.

In November 2013, India launched a Mars orbiter named Mangalyaan to map potential sources of methane plumes which might indicate the presence of a microbe biosphere deep beneath the Martian surface. When asked why the country would invest in such costly programs, Nisha Agrawal, CEO of Oxfam—a confederation of charitable organizations focused on alleviating poverty—told BBC:

> *India is home to poor people but it's also an emerging economy, it's a middle-income country, it's a member of the G20. What is hard for people to get their head around is that we are home to poverty but also a global power...We are not really one country but two in one. And we need to do both things: contribute to global knowledge as well as take*

care of poor people at home.

K. Radhakrishnan, chair of the Indian Space Research Organization (ISRO), elaborated:

> *Why India has to be in the space program is a question that has been asked over the last 50 years. The answer then, and now and in the future will be: 'It is for finding solutions to the problems of man and society.'*[cc]

President Xi Jinping has made it very clear he intends to have China establish itself as a space superpower.

China sent its first astronaut into space in 2003, the third country after Russia and the United States to achieve independent manned space travel. In 2013, three Chinese astronauts spent 15 days in orbit and docked with an experimental laboratory as part of Beijing's plan to establish an operational space station by 2022.

The first of three 20-ton modules are scheduled for launch to the station in 2020. The core module, called "Tianhe" (Harmony of Heaven), will house three astronauts and carry their supplies for a stay of several months. Over the following two years, two laboratories for scientific equipment will be added.

As Chinese Ambassador to the United Nations Shi Zhongjun explained:

> *The Chinese Space Station belongs not only to China, but also to the world...Guided by the idea of a shared future, the [Chinese Space Station] will become a common home for all humankind. It will be a home that is inclusive*

and open to cooperation with all countries…

At 36, teams from around the world have already applied to send experiments to the station.

China is opening up its manned and lunar space programs to international participation with the focus on nations that have not had access to space technology in the past. Originally centered on cooperation with neighbors in Asia, the infrastructure, technology and cultural exchanges with more than 100 nations now span the world from Asia to Europe and Africa.

The Chinese government has made it clear that they are very committed to human lunar and planetary exploration. In 2013 they launched a robotic rover called "Yulu" (Jade Rabbit) that surveyed for useful natural resources in the northwest corner of the giant Imbrium Basin, the left eye of the "Man in the Moon." Zhao Xiaojin, director of aerospace for the China Aerospace Science and Technology Corporation, described the rover as "a high altitude patrolman carrying the dreams of Asia."

The next stage to follow will likely land a lunar probe, release a Moon rover and return a probe to Earth.

In December 2018, China became the first country to land a spacecraft on the far side of the Moon.

Space affords a natural supply depot stocked with a vast assortment and quantity of potentially useful, even critical, materials that can expand human experience and enterprise. Of these rich resource caches, the Moon is the closest, representing a relatively near-term source and laboratory for rocket propellant, oxygen and water production and an operational base for development and demonstration of other extraterrestrial technologies.

In 2018, an International Space Exploration Coordination

Group (ISECG) comprised of space agencies representing numerous countries released a *Global Exploration Roadmap* which explains why space exploration is a global priority:

> *Space exploration offers a unique and evolving perspective on humanity's place in the Universe. It stimulates curiosity and the ability to see the bigger picture. By uncovering new information about the beginnings of our Solar System, space exploration brings us closer to answering profound questions that have been asked for millennia: What is the physical nature of the Universe? Is the destiny of humankind bound to Earth? Are we and our planet unique? Is there life elsewhere in the Universe?*[cci]

As summarized in the same report:

> *ISECG space agencies envision that by the mid-2020s a Gateway in the lunar vicinity will open the space frontier for human exploration of the Moon, Mars and asteroids as we expand human exploration and commerce into deep space.*

The ISECG proposal goes on to explain:

> *The Gateway will support activities on and around the Moon while also serving as a technology and operations testbed allowing humans to address the challenges and risks of deep space exploration and conduct scientific*

Larry Bell

investigation of our Solar System.

ISECG partners would collaborate to develop advanced technologies that will carry humankind to the Moon, Mars and beyond. As highlighted:

> *A partnership between humans and robots is essential to the success of this venture. Robotic missions accomplish world-class science while also serving as our scouts and proxies, venturing first to hostile environments to gather critical information that makes human exploration safer. Humans will bring their flexibility, adaptability, experience, dexterity, creativity, intuition, and the ability to make real-time decisions to the missions.*

Buzz Aldrin believes that the Cold War space race during the 1960s and '70s to outperform the former Soviet Union has reached the finish line:

> *In Korea, we knew we were really fighting the Soviets as well as the North Koreans, and a strong sense of competition on our part carried into the space race. We were determined not to let the "Ruskies" beat us in Korea, and we certainly weren't going to let them get the upper hand in space.*
> *A second race to the Moon would be a waste of precious resources. It will offer no unique American glory or payoffs in either commercial or scientific terms. It's high time now to raise our vision and commitments to*

212

loftier, more far-reaching goals in global cooperation.

Apollo was all about a get-there-in-a hurry straightforward space race strategy and don't waste time developing reusability. That chapter in the space exploration history books is closed. Instead, I urge that all spacefaring nations join a unified international effort to explore and utilize the Moon through a partnership that involves commercial enterprises and other nations.

Prospecting for Resourceful Opportunities

The Moon, our closest celestial neighbor, offers a variety of interesting scientific features and potentially valuable resources.

Recent discoveries of large quantities of water on the Moon have excited great interest for a multitude of applications which can greatly reduce dependence upon costly transportation of Earth-delivered consumables. The precious molecule is vital to life for drinking and constitutes a source of oxygen for breathing. It's also a source of hydrogen and oxygen for rocket propellant to fuel ascent vehicles from the lunar surface and transportation to cis-lunar (Moon-Earth-orbit) space and beyond. It can also provide highly mass-efficient solar and nuclear radiation shielding material for astronauts.

Orbiting Moon-circling instruments reveal that the Moon's North Pole alone may contain as much as 600 metric tons of ice. Converted to liquid hydrogen and liquid oxygen fuel, this lunar-derived surface water can support human activities and provide a source of fuel for Mars-bound rockets.

Another resource of special interest is helium-3 deposited on the lunar surface by solar wind which might serve as a valuable fuel for future fusion power. If world fusion technology proves successful, it is estimated that each ton of the material would yield energy equal to approximately 50 million barrels of crude oil. My friend Apollo 17 astronaut Harrison Schmitt, the last human to step on the Moon's surface, is a strong advocate to fuel electricity both for Earth and for bases and operations on the Moon and Mars.

Extended distance and time operations on human space voyages and lunar/planetary surfaces will require that means are afforded to harvest and use water and other extraterrestrial resources for propulsion, life support and perhaps eventually, for construction and even commercial export.

Findings obtained from NASA's Mars Reconnaissance Orbiter (MRO) offer evidence that water also exists on Mars, and that some even intermittently flows on its surface. MRO's imaging spectrometer detected darkish streaks up to a few hundred meters in length along with spectral signatures of hydrated minerals which darken and extend down deep slopes during warmer seasons (above minus 10 degrees Fahrenheit).

Since these features fade and eventually disappear altogether in cooler seasons, it is theorized that hydrated salts lower the freezing point of the liquid brine, just as salt causes ice and snow on roads here on Earth to melt more rapidly. However, it's presently unclear whether the dark streaks are signatures of the salts, or rather, appear as a result of the existence of briny water that periodically wicks up from subsurface sources.

Mars also has a great diversity of other resources, some of which are present and most likely more accessible than on the Moon. For example, carbon dioxide, nitrogen and hydrogen exist on the Moon, but unlike Mars, only in tiny parts per million quantities.

Mars isn't known to have helium-3, but does have an abundance of deuterium, a heavy isotope of hydrogen that could be used for future nuclear fusion. And while oxygen is abundant in lunar soil, it is tightly-bound in oxides such as silicon dioxide (SIO_2), magnesium oxide (MgO), ferrous oxide (Fe_2O_3) and aluminum oxide (Al_2O_3) which require high energy processes to release.

The Cebrenia quadrangle site located in the northeastern portion of Mars has been shown by the Mars Orbiter Laser Altimeter satellite to have water ice about four inches below the surface along with soil rich in silicon, iron, magnesium, sulphur, calcium and titanium.

While robotic exploration of Mars has yielded tantalizing clues about what was once a water-soaked planet and has revealed frozen water still trapped below the surface, the best way to study Mars is with the two hands, two eyes and two ears of a geologist.

Humans surpass machines in speed, efficiency, nimbleness and the dexterity to go places and do things. Unlike machines, we have the innate smarts, ingenuity and adaptability to evaluate and respond in real-time situations...to improvise, and to prevail over surprises.

Elevating Pioneering Aspirations and Possibilities

If rising above such daunting challenges seem truly "remote," consider how leading "authorities" viewed notions of heavier-than-air flying machines prior to the Wright Brothers' first flight in 1903 and Robert Goddard's predictions less than two decades later that rockets could reach the Moon, much less that humans would walk on its surface little more than a half-century later.

Astronaut Buzz Aldrin urges that it is once again time to dare to pursue big dreams:

Although space exploration progress has slowed, it is my great hope that a new generation of leaders and doers will once again boldly venture where no one has gone before. Our Apollo days were a time when we did bold things, achieving leadership. Now is our time to be bold again in space.

Here, humankind's primary spacefaring destination should be the Red Planet:

Mars represents a new world of opportunity and discovery. Scientific and public interest in the planet has grown since 1960 telescope-driven observations have since been augmented by voyages of numbers of automated spacecraft sent there by multiple nations. It has been flown by, orbited, smacked into, radar-examined, rocketed onto, bounced upon, rolled over, shoveled, drilled into, baked, and even laser-blasted. Still to come: Mars being stepped upon. The first footfalls will mark a historic milestone.

Permanent Mars habitation such as is exemplified in the popular movie *"The Martian"* will require preparations for long-term survival far beyond Earth lifelines that pose a variety of enormously challenging innovation tests. The new pilgrims will require an ability to live off the land, a circumstance that 102 other adventuresome souls once bravely faced upon leaving England for a New World aboard the Mayflower voyage. Martian settlers, however, will face much stiffer challenges. As illustrated in the movie, a crop failure could bring disastrous consequences.

Buzz clearly has strong personal experience to recognize the great human dangers and demands:

> *Inevitably, Mars settlement will invoke risks and casualties, just as other pioneering ventures have. Unfortunately, pioneers will always pave the way with sacrifices. Over the decades, we have lost numbers of individuals—several of them close personal friends of mine—all intent on pushing the boundaries of exploration and seeking new horizons. Risk and reward is the weighing scale of exploring and taming space.*
>
> *Successful innovators and doers who conceive things and make them happen combine awareness of failure risks of worthwhile enterprises with the willingness to take them. They are patient, resilient and don't quit. They experiment, often fail, learn more, and start over. Great companies like Google and Apple provide a culture that empowers employees to stretch their curiosity and creativity to explore "impossible dreams" that often don't work out.*

As Buzz also observes in his book *No Dream is Too High,*

> *Failure is not a sign of weakness. It is evidence that you are alive and accepting of worthwhile risks.*[ccii]

A NASA report concludes that "a strong motivating factor for the exploration of Mars is the search for extraterrestrial life," but maybe what we will discover there is us. Maybe what we will discover are exceptional human potentials we can now only dream about.

EXPERIENCING EXCEPTIONAL LIVES

WHILE WE'RE AT it, we can also strive to live our own lives more boldly and fully.

American science fiction writer Robert Heinlein has proposed an ambitiously comprehensive range of expectations regarding what doing so might mean in his book *Time Enough for Love*:

> *A human being should be able to change a diaper, plan an invasion, butcher a hog, conn a ship, design a building, write a sonnet, balance accounts, build a wall, set a bone, comfort the dying, take orders, give orders, cooperate, act alone, solve equations, analyze a new problem, program a computer, cook a tasty meal, fight efficiently, die gallantly. Specialization is for insects.*[cciii]

Or as British-born neurologist and science writer Oliver Sacks instructed in a 2012 *New Yorker* magazine article, living boldly and fully requires conscious contemplation of greater life and world potentials:

> *To live on a day-to-day basis is insufficient for human beings; we need to transcend, transport, escape; we need meaning, understanding, and explanation; we need to see over-all patterns in our lives. We need hope, the sense of a future. And we need freedom (or, at least, the illusion of freedom) to get beyond ourselves, whether with telescopes and microscopes and our ever-burgeoning technology, or in states of mind that allow us to travel to other worlds, to rise above our immediate surroundings.*[cciv]

Sacks further observes that such consciousness often requires peaceful minds:

> *We may seek, too, a relaxing of inhibitions that makes it easier to bond with each other, or transports that make our consciousness of time and mortality easier to bear. We seek a holiday from our inner and outer restrictions, a more intense sense of the here and now, the beauty and value of the world we live in.*

The ultimate benefit of thinking and living whole is realized through expanded and strengthened connections to life experiences. Writing in *The New Yorker*, David Brooks summed this up eloquently:

> *I've come to think that flourishing consists of*

putting yourself in situations in which you lose self-consciousness and become fused with other people, experiences, or tasks. It happens sometimes when you are lost in a hard challenge, or when an artist or a craftsman becomes one with the brush or the tool. It happens sometimes while you're playing sports, or listening to music or lost in a story, or to some people when they feel enveloped by God's love. And it happens most when we connect with other people.

Brooks concludes:

I've come to think that happiness isn't really produced by conscious accomplishments. Happiness is a measure of how thickly the unconscious parts of our minds are intertwined with other people and with activities. Happiness is determined by how much information and affection flows through us covertly every day and year.[ccv]

As discussed in my earlier book, *Thinking Whole: Rejecting Half-Witted Left & Right Brain Limitations,* living life to the fullest is evidenced in everyday experiences through our awareness of surrounding environments. It is expressed through curiosity which compels our interest in how and why natural and man-made things work the way they do...interconnected relationships between ourselves and others...patterns and rhythms observed in nature...spiritual lessons and explorations that motivate higher purposes and values...inspirations experienced through image forms, literature and stories of the past...music...everything combined that our whole minds can contemplate.

Thinking and living whole involves tapping into our natural personal resources. This entails following passions; seeking out curiosities with drive and vigor; employing powers of intuition; and letting our minds flow with free thinking, boldness and discipline to bring worthwhile ideas to fruition.

Using your whole brain doesn't necessarily qualify you for modern-day "Renaissance man" (or woman) distinction, although it may give you a head start. After all, it's a bit daunting to live up to the romanticized notion of someone who can do just about everything better than almost everyone else.

Today, such multi-talented individuals frequently referred to as "polymaths" are characterized as possessing a broad spectrum of interests, proficiencies and achievements; being curious and eager to learn about a variety of topics; focusing on original and creative problem-solving; and holding abilities as talented communicators who make complex principles and ideas accessible to others.

As former British Prime Minister Winston Churchill, a polymath who won Nobel Prize for Literature in 1953, reflected:

> *If it weren't for painting, I wouldn't live. I wouldn't bear the extra strain of things.*

Being a polymath doesn't require genius. It does, however, challenge us to constantly open and stretch our whole-mind connections with natural talents developed and applied as broadly as possible.

In his 1953 essay *The Hedgehog and the Fox,* Oxford University social and political philosopher Isiah Berlin presents a hedgehog who knows a lot about a single, narrow subject, and a fox who knows a little about many subjects. While both of the critter mindsets are deficient, populations that operate with a "hedgehog mindset" appear to be gaining.[ccvi]

Psychologist Robert Plomin, a professor of behavioral genetics at King's College, London, observes:

> *Nowadays the training is so specialized, But the big advances come from the foxes who know a little bit about a lot of things and can put two and two together, rather than the hedgehogs in the trenches who are burrowing away and trying to find out more and more about less.*[ccvii]

I have found this to trend to be particularly true in academia.

Exercising Conscious Awareness and Curiosity

This remarkable ability of our minds to discover and recognize distinct patterns enables us to consider, sort out, prioritize and act upon complex sets of inputs containing hundreds of features based upon comparisons of lifetimes of recalled lessons. In her *New York Times* essay *I Sing the Body's Pattern Recognition Machine*, Diane Ackerman points out that just as it keeps us out of trouble, pattern recognition also pleases us, "rewarding minds seduced, yet exhausted by complexity:"

> *We crave pattern, and find it all around us, in petals, sand dunes, pine cones, contrails. Our buildings, our symphonies, our clothing, our societies—all declare patterns. Even our actions: habits, rules, codes of honor, sports,*

traditions—we have many names for patterns of conduct. They reassure us that life is orderly.

As children, we learn subtle patterns from our parents, including the texture of their senses and their emotional style. Just as we learn the alphabet and that teeth can bite—horse teeth or brother teeth—we learn the configurations of cuddling, the emotional contours of mother's voice, the silhouette of a friend.

We rely on patterns, but we also cherish and admire them. Few things are as beautiful as a ripple, a spiral or a rosette. They are visually succulent. The mind savors them. Societies like to invent patterns of action, rules to cushion nature's laws. And word patterns: Madam, I'm Adam. Patterns reflect one of the brain's deepest needs—to fill the world with pathways and our lives with a design.[ccviii]

Ackerman warns us to remember, however, that patterns we think we recognize can also be misinterpreted because "generalizing, even from concrete details, isn't always accurate. You were wrong. It's not your mother. The woman across the street only bears a resemblance to her. She's not waving to you, but to the person walking behind you."

Recognizing stored patterns from past learning experiences is routinely useful when we apply those lessons to logical conceptualization and problem-solving activities. Our brains do this using a combination of qualitative rules, heuristic processes and more well-defined deterministic assessments to discover and match previous observations to present conditions. An analogy might compare entirely logic-based chess-playing computers in competition with human expert-level players who depend heavily upon patterns catalogued from former winning and losing moves.[ccix]

Complex problem-solving typically involves mental attempts to balance holistic "big picture" thinking and reductionist thinking with the fullest possible consideration of all known component parts, relevant contributing influences, and linkages between elements that comprise the entire system.

This broad idea is often credited with building upon several influential philosophical and scientific roots. Included are discussions of "holism" advanced by South African statesman Jan Smuts in the 1920s, general systems theory proposed by Ludwig von Betalanffy in the 1940s and cybernetics concepts presented by Ross Ashby in the 1950s along with works by Jay Forrester at MIT in the field of system dynamics.[ccx],[ccxi],[ccxii],[ccxiii]

Broadly described, the general system theory, or "general science of wholeness," recognizes as a truism that "the whole is more than the sum of its parts." It follows, then, that characteristics of a complex "thing" are not explainable from characteristics of isolated parts alone.

As Ludwig von Betalanffy explained, a system is a complex of interacting elements that are open to and interact with their environment and which can acquire qualitatively new properties through emergence. Accordingly, they are generally self-regulating (self-correcting through feed-backs) in a continuing state of evolution.

Thinking holistically encourages us to observe our world, surroundings, issues and connections from larger and more dynamic perspectives before delving into details... a classic lesson of viewing a forest before becoming lost in trees and entangled in underbrush. In doing so, we may avoid recurring problems made even worse by failed attempts to fix the wrong parts. A big picture perspective may also reveal that a "best solution" may be to recognize and exit bad environments that pose undesired risks.

Looking at broader pictures raises and expands abilities to conceive new possibilities. At the same time, it's also important to be able to shift awareness to the smaller component elements. That's a big part of holistic thinking as well.

Much of our mental processing applies "reductionist thinking," which removes outside influences considered as distractions in order to focus attention upon aspects which matter most. Applied to biological and social systems, for example, this often involves removing or amplifying certain outside stimuli to determine which behavioral influences are of greatest and least importance.

The "scientific method" fundamentally employs a reductionist process, one which has led to pioneering of incredible advances that have made our lives healthier, more comfortable and more productive. The trick is to not only understand which parts of a system to isolate and examine, but also to figure out why they interact together in certain ways and under certain conditions. Attempts to optimize certain of those parts without regard to their overall connections can tamper with the success of the system and the functioning and wellbeing of its surrounding environment.

Consider, for example, the many side effects on warning disclaimers attached to common pharmaceuticals. Pesticides that kill one pesky critter can result in infestations by others which are potentially worse.

Similar methodological problems arise in understanding how the arrangements and interactions of special groups of molecular components which interact as neurons along with hormones and numerous other bodies and systems enable consciousness and logical analyses to better understand how we, altogether—as thinking, moving, changing, self-correcting, procreating systems—"work."

A rather recently emerging goal-results-oriented "integrative thinking" analytical methodology begins by defining a problem to be solved as the difference between a current circumstance and what one wishes to achieve. Its credited originator, Graham Douglas, describes this approach as a process of integrating intuition, reason and imagination in a human mind with a view to developing a holistic continuum of strategy, tactics, action, review and evaluation for addressing a problem in any field as applied.

Originated by Roger Martin, Dean of the Rotman School of Management at the University of Toronto, and colleague Mihnea Moldoveanu, Director of the Desautels Center for Integrative Thinking, the teachable discipline is designed to facilitate associations between known internal and external factors which may have previously not been recognized.

Integrative thinkers are characterized to differ from conventional thinkers in a variety of dimensions. For example, rather than seeking to simplify a problem as much as possible, they instead consider most variables to be salient and focus special attention upon alternative views and contradictory data. In doing so, they embrace a complex understanding of how causality regarding salient features interconnect and interact in multi-directional ways, rather than limiting relationships according to linear, one-way dynamics.

When faced with two opposing solutions forcing trade-off evaluations, integrative thinkers strive for creative resolutions of tension rather than accepting simple choices. The Rotman School of Management website explains this as: "generating a creative resolution of the tension in the form of a new model that contains elements of the individual models, but is superior to each."

So, what makes a big picture "big," and some ideas larger than others? Although there are, of course, nearly endless circumstances calling for subjective assessments, here are a few considerations.

We might imagine that an idea is big if it helps us sort out ways to make more sense out of confusing conditions, influences and interrelationships in a complex field of events or systems. It's big if it enables us to identify missing parts of a puzzle and recognize how they fit into a coherent picture...to meaningfully connect what previously appeared to be random patterns of fragmentary dots...to provide hierarchical logic structures and guidelines that organize and prioritize information in more useful ways...to hypothesize testable theories that change and illuminate understanding of important principles and issues...An idea is as big as each of us cares to value it.

Passionate dedication to high ideals and purposes can inspire and challenge us to set goals and expectations that reach farther, aim higher and think better. Robert Heinlein observes that pursuing passions stimulates our curiosity, inciting us to courses of inquiry and creative action that reveal unexpected possibilities. We should appreciate them, nurture them and enjoy them.[ccxiv]

Leo Burnett, founder of the Leo Burnett Worldwide Inc. advertising company, has observed, as I also have:

Curiosity about life in all of its aspects, I think,
is still the secret of great creative people.

Albert Einstein, whose revolutionary theory of general relativity exemplified holistic thinking, purportedly said:

I have no special talent. I am only passionately

curious." Nor was science his only passionate interest.

Einstein is also broadly quoted as saying:

If I were not a physicist, I would probably be a musician. I often think in music. I live my daydreams in music. I see my life in terms of music...I get most joy in life out of music.

American vocalist Judy Collins shares that:

I think people who are creative are the luckiest people on Earth. I know that there are no shortcuts, but you must keep your faith in something Greater than You, and keep doing what you love, and you will find the way to get it out of the world.

Discovering Personal Creativity

Einstein attributed his creativity as a benefit of curiosity resulting from a "long childhood."

As a child, Einstein was slow in learning how to talk. He exhibited evidence of a condition known as "echolalia," and possibly a mild form of autism or Asperger's syndrome. Whenever he had something to say, he would first softly whisper it to himself two or three times until it sounded good enough to pronounce out loud. That quirk prompted his family maid to refer to him as "der Depperts" (the dopey one). Some relatives described him as "almost backwards." [ccxv]

Einstein believed that his slow verbal development later served as an advantage, allowing him to observe with wonder the everyday phenomena others took for granted. He later reflected:

> *When I ask myself how it happened that I in particular discovered the relativity theory, it seemed to lie in the following circumstance. The ordinary adult never bothers his head about the problems of space and time. These are things he has thought as a child. But I developed so slowly that I began to wonder about space and time only when I was already grown up. Consequently, I probed more deeply into the problem than an ordinary child would have.* [ccxvi],[ccxvii]

Einstein retained the curiosity and awe of a child throughout his life. He wrote to a friend: "People like you and me never grow old. We never cease to be like curious children before the great mystery into which we were born." [ccxviii]

Young Einstein, however, was far more than an ordinary child. He recalled his father's gift to him of a magnetic compass at the age of four or five as such a profound "great awakening experience" that he trembled and grew cold. He marveled that the needle behaved as if influenced by mysterious powers of some hidden force, manifesting a sense of wonder that motivated him throughout his life.

As he later reminisced, "I can still remember—or at least I believe I can remember—that this experience made a deep impression on me. Something deeply hidden had to be behind things." [ccxix]

Walter Isaacson reminds his readers that the magnetic compass experience may be correlated with Einstein's life-long interest in field theories to describe nature. In a gravitational or electromagnetic field there are forces that can act on a particle at any point, and equations of field theory describe how these change as one moves through the region. His Theory of General Relativity is based upon equations that describe a gravitational field. Throughout his life he fervently pursued hope that such field equations would form a basis for a "Theory of Everything."

Young Einstein exhibited abundant imagination, confidence and tenacity that persisted throughout his life. Above all, he believed "Imagination is more important than knowledge."

Albert Einstein was perpetually drawn to contemplate and seek simpler solutions to complex puzzles and was willing to challenge orthodox assumptions that others took for granted. His sister, Maja, remembered that he persistently succeeded in building houses of cards up to 14 levels tall.

He reported in a 1935 interview:

> *As a boy of 12, I was thrilled to see what is possible to find out truth by reasoning alone, without the help of any outside experience. I became more and more convinced that nature could be understood as a relatively simple mathematical structure.*[ccxx]

That early insight paid big dividends. In 1895, 16-year-old Einstein imagined what it would be like to ride alongside a light beam. A decade later, this boyhood musing provided the conceptual foundation for two great advances of 20th century physics: relativity and quantum theory. Then in 1915, only one more decade after that, he followed that light beam of imagination to produce his everlasting crowning scientific accomplishment...the General Theory of Relativity.

Nikola Tesla also recognized the importance of nurturing youthful imagination. He wrote in his autobiography:

> *Our first endeavors are purely instinctive,
> promptings of an imagination vivid and
> undisciplined. As we grow older, reason asserts
> itself and we become more and more
> systematic and designing. But those early
> impulses, although not immediately
> productive, are the greatest moment and may
> shape our very destinies. Indeed, I feel now
> that had I understood and instead cultivated
> instead of suppressing them, I would have
> added substantial value to my bequest to the
> world. But not until I had attained manhood
> did I realize that I was an inventor.* [ccxxi]

As Albert von Szent-Gyorgi, the Hungarian Nobel Prize-winning physiologist who first discovered the benefits of vitamin C, was fond of saying, "Discovery lies in seeing what everyone sees, but thinking what no one else has thought."

Such imaginative "creativity" defies boundaries of any singular scientific phenomena or authoritative perspective. As described in the blog *Psychology Wiki,* it has been "attributed variously to divine intervention, cognitive processes, the social environment, personality traits, and chance ('accident or serendipity'), genius, mental illness and humor." [ccxxii]

Nevertheless, however we define it, true creativity cannot occur without whole brain mindfulness. Anything less would invariably produce only half-baked ideas. It would also be a terrible waste. As Thomas Edison, an indisputably creative fellow reportedly remarked, "The chief function of the body is to carry the brain around."

From a spiritual perspective we might think of creativity as the product of imagination put to purposeful ends of the creator. Although exactly how that happens is as a highly individual and subjectively evaluated matter, it is clearly a topic of interests that sells many books and other print publications. I can therefore only wish that this one will be included...although hopefully not according to a genre category described by Stephen Asma, author of *The Evolution of Imagination*.

Chicago Columbia College professor Asma observes:

> *Books about creativity have tended to fall into one of three genres. On the one hand, there have been breathless and over-reaching feel-good paeans to famous entrepreneurs and successful CEO creatives. This kind of book is crammed with amusing but shallow factoids and over-interpreted fMRI studies, all wrapped up in a vaguely inspirational glaze.*

"Next," he continues:

> *We have the how-to books that give artists a series of exercises to unblock their creative flow. These books are either therapeutic or instructive, or both, and seek to nurture the joy of our inner prodigy.*

Asma concludes:

> *The third genre is the impenetrable academic buffer, chock-full of erudite and cryptic references to Foucault and the hegemonic phallocentric horizon of being, but otherwise devoid of illumination.*[ccxxiii]

Our most productive and innovative thinking often seems

effortless. We happily engage in this for many hours at a time, doing what we truly care about and enjoy without consciously "thinking" or using "willpower" at all.

Philosopher Stephen Asma, also an improvisational jazz musician who has played with such notables as Miles Davis, Bo Diddley, Buddy Guy and Otis Rush, points out a need for creative artists to be able to clear their minds in order to get "in the zone." This involves finding ways to enter "Zen moments" of being fully present in an egoless state. He notes that:

> *Buddhist artists have been doing this for thousands of years, producing amazing poetry, ink drawings, calligraphy, bonsai, ikebana, sculpture, and so on.*[ccxxiv]

Asma acknowledges that this egoless state also presents a paradox. He observes that in order to be fully in the "here and now:"

> *I must actually shut off all imaginative creation of the future (what could be), and the re-creation of the past (what was), as well as the imaginative possibilities of the alternative present.*

He continues:

> *The present moment is a singularly unimaginative place. In Chan (Zen) Buddhism, the artist celebrates this empty (no self) moment, even as she drags ink across rice paper. In contrast, Tibetan Buddhism celebrates a creative visualization tradition that*

is deeply imaginative and filled with rich narratives and images about past, present, and future beings of various power and influence.

Other psychologists of creativity, like Mihaly Csikzentmihalyi, have also glorified this condition of creativity as "flow...a state of effortless concentration so deep that they lose their sense of time, of themselves, of their problems."

According to Csikzentmihalyi, these experiences occur when there is a balance between challenges and skills, and when action and awareness merge, self-consciousness disappears, time becomes distorted and the activity becomes an end in itself with inherent rewards. Descriptions of the joy expressed by people in this state are so compelling, that Czikszentmihalyi calls it an "optimal experience." [ccxxv],[ccxxvi]

In this true meditative state, the past is already gone, and thinking about the future becomes a subjective exercise in imagining nonexistent events. Nevertheless, we also live in co-present simultaneous possible worlds made up of "almosts" and "what ifs" and "maybes" which are in various processes of happening. Imagination expressed through intuition is required to help us prepare for them too.

So maybe try to relax. Take at least an occasional break to vacation in that state of bliss where time zones, self-consciousness and problems cease to exist.

As philosophy professor Asma both asks and answers:

Why do we have imagination? One major answer is functional and utilitarian...evolutionary adaptation, for example. The other is because it provides some of the highest human pleasure and joy. The fantasy view of imagination tilts in this

*direction (though fantasy can also be recruited
for adaptive survival ends).*[ccxxvii]

Immanuel Kant and Aristotle viewed the primary role of
imagination as an unconscious faculty that gathers and
synthesizes sensory perceptions into coherent and universally
applicable representations.

Kant tended to regard imagination primarily as a
synthesizer of sensibility and understanding...as a form of
judgment rather than one of fantasy or creativity. In his treatise
Poetics, Aristotle praised artistic forms of imagination for
shaping a version of real events such that a higher truth
emerged.[ccxxviii]

British naturalist Charles Darwin saw imagination as a
faculty that created brilliant and novel results by uniting former
images and ideas. Applying this to his personal life, he reflected:

> *I attain the highest level of adaptive
> imagination when I have voluntary control
> over the uniting impressions and scenarios—
> when I can conduct internal simulations of
> possible outcomes (using impressions, folk
> physics, and variable conditions).*[ccxxix]

Darwin wrote in his famous work, *Descent of Man*:

> *The value of the products of our imagination
> depends of course on the number, accuracy
> and clearness of our impressions, on our
> judgement and taste in selecting or rejecting
> the involuntary combinations, and to a certain
> extent on our power of voluntarily combining
> them.*[ccxxx]

Viennese psychiatrist Sigmund Freud envisioned imagination hard at work in the dark poetics of dreams as a necessary release system for antisocial desires. And while it may be tantalizingly seductive to occasionally attempt to correlate and interpret particularly provocative dreams with triggering causes and profound "meanings," such mental theater is often likely to reflect more nonsense than neuroses.

As Stephen Asma points out, our dreams are frequently more emotional than intellectual. Unlike more logical, sequential and linear contemplation in our waking life, they represent highly intuitive and image-based forms of thinking and feeling. Since dreams do not signal to the subject whether the experience is real or a figment—the senses can frequently be tricked.[ccxxxi]

Asma characterizes dreams as "improvisations in the sense that they are autonomous, uncontrolled narratives with loose cause-and-effect sequencing. In fact, some dreams may only be 'brain noise'—the hum and flicker of a big wet machine in rest phase." During deep rapid eye movement (REM) periods of sleep, storms of neuronal firing sweep through the brain, while the neural systems most active during awakened times cease firing completely.

The mentally restorative quality of sleep we experience may be influenced by our individual levels of calming and relaxing serotonin. Many mathematicians and scientists report getting their "aha moments" after they have relaxed their conscious pursuit of a solution. When the problem sinks down into the unconscious, it continues to have a life, as it were—a private life that consciousness is not privy to.[ccxxxii]

Aristotle described imagination as "a faculty in humans (and most other animals) that produces, stores, and recalls the images used in a variety of cognitive and volitional activities." This faculty serves as a driving force for smarter, non-cliché

237

thinking in a great variety of human endeavors. Included are the visual, literary and performing arts; science and engineering; marketing and economics; and ethics and government. While often annotated with lightbulb metaphors, lightning flashes, sparks and other combustion symbols, many of those insights quietly emerge without drama or fanfare.

Stephen Asma observes that, as with Plato and Aristotle, many philosophers throughout history intentionally or inadvertently demote "imagination" to become a sort of "weak knowledge," making it derivative or secondary to "real knowledge." In doing so, they have tended to think of real knowledge as "a process of seeing through the particular cases to the universal rules or laws that govern them."

Asma describes imagination from a whole-brain sensory perspective as "an embodied voluntary simulation system that draws upon perceptual, affective, and memory elements, for the purpose of creating works that adaptively investigate external and internal resources."

In addition to being an extrinsically useful "adaptive investigation system," imagination also possesses a capacity to experience significant intrinsic value in joy of play and states of wonder. This second system accomplishes its synthetic work through mechanisms which are distributed across various regions that access and control cognition and modes of communication.

Imaginative play enables us to take ideas "off-line" and rehearse them before taking action. Einstein reputedly said that "play is the highest form of research," claiming that his mind engaged in a kind of "combinatory play" or "associative play" just before his major breakthroughs.

His local analysis would follow after this synthesizing creative phase.[ccxxxiii]

Einstein also said:

I am not enough of an artist to draw freely upon my imagination. Imagination is more important than knowledge. Knowledge is limited. Imagination encircles the world.

Or as Harvard psychiatrist Arnold H. Modell puts it, our minds have the ability to create a "second universe"—an internal environment of possibilities that exists concurrently with the stubborn physical world.[ccxxxiv]

Our imaginative powers increase as we repeatedly exert voluntary control over many downstream modes of cognition and culture (e.g., through technology, innovation, storytelling, music, etc.). These events involve the manipulation of information-rich perception, memories, image schemas and bodily gestures born out of emotional needs and social experiences.

Here, our evolved present-day human intellect can be viewed as both a product and servant of our social life. Our improvising imagination—our early intellect—gave us the behavioral and mental scaffolding to organize and manage our experiences. In fact, this began to occur long before human imagination invented language and word concepts.

Modern culture owes its very existence to imaginative innovation, a uniquely human cognitive process that synthesizes and acts upon observations, musings and theories to produce new or modified insights and products. As da Vinci, Einstein, Edison and countless other scientific and technical theorists, inventors and practitioners have demonstrated, innovation born of imagination is no longer viewed as the exclusive domain of the fine arts.

Asma points out that this imaginative process involves the whole brain. He writes:

Larry Bell

> *There is no imagination organ buried in the
> neuroanatomical structures of the brain.
> Several candidates for location have come and
> gone, most popular of which is the idea that
> the right hemisphere houses imagination. But
> data suggest no clear localization of creativity,
> and the most that can be said with confidence
> is that communication between brain regions
> is very high in imaginative people.*[ccxxxv]

There may be a solid scientific basis to defend "scatter-brained ideas" after all. Those mental communications interconnect information processing centers which are distributed throughout many locations. That creative network includes centers within the emotional brain (limbic system), memory system and motor system that interact with the rational brain (neocortical deliberation system).

As David Eagleman points out, "We are masters at generating alternative realities, taking what is and transforming it into a panoply of what-ifs." He attributes much of this capability to what he terms a "creative economy" of thinking which catalogs and stores important past lessons so that we don't need to devote needless energy relearning them.[ccxxxvi]

Eagleman observes that navigating the human world is a difficult and energy-expensive endeavor that requires moving around and using a lot of brainpower. It conserves a lot of that energy when we can make correct short-cut predictions:

> *When you know that edible bugs can be found
> beneath certain types of rocks, it saves turning
> over all rocks. The better we predict, the less
> energy it costs us. Repetition makes us more
> confident in our forecasts and more efficient in*

our actions.[ccxxxvii]

Repetition also presents downside problems when we allow creative thinking to ebb. David Eagleman and his co-author Anthony Brandt remind us that too much familiarity through repetition often breeds indifference. Waning attention causes us to put less effort into understanding something better, exploring new ideas and weighing alternative solutions. They note: "This is why marriage needs to be constantly rekindled. This is why you'll only laugh so many times at the same joke."

Our innovative spirit introduces surprises that keep us awake to new experiences and possibilities. Surprise engages us. Surprise allows us to escape autopilot. Surprise gratifies us.

All of this requires our minds to conduct perpetual balancing acts. As Eagleman and Brandt describe the situation:

> *On one hand, brains try to save energy by predicting away the world; on the other hand, they seek the intoxication of surprise. We don't want to live in an infinite loop, but we also don't want to be surprised all the time.*[ccxxxviii]

International business advisor and author Margaret Hefferman attributes great creative importance to pattern recognition...most particularly in her corporate field, as an ability to discern patterns in tons of data. She observes:

> *Your mind collects that data by taking note of random details and anomalies easily seen every day: quirks and changes that, eventually, add up to insights.*

241

Venezuelan-American filmmaker and public speaker Jason Silva agrees, observing:

> *Creativity and insight almost always involve an experience of pattern recognition; the Eureka moment in which we perceive the interconnection between disparate concepts or ideas to reveal something new.*

In her book *Breakthrough Creativity: Achieving Top Performance Using the Eight Creative Talents,* Lynne C. Levesque emphasizes that creative people are unafraid to challenge the status quo:

> *To be creative you have to contribute something different from what you've done before. Your results need not be original to the world; few results truly meet that criterion. In fact, most results are built on the work of others.*

Creativity springs much less a desire to be different than from a willingness to be different in pursuit of something fresh or better. The well-known writer, reporter and political commentator Walter Lippman famously quipped:

> *When all think alike, then no one is thinking.*

As the late American jazz composer and bandleader Charles Mingus pointed out:

> *Anybody can plan weird; that's easy. What's harder is to be simple as Bach. Making the*

simple awesomely simple, that's creativity.

German-American architect Helmut Jahn also expressed this quest for essential simplicity, stating:

> *Creativity has more to do with the elimination of the inessential than with inventing something new.*

Developing and Trusting Intuition

Our intuition draws upon memories and biases gained through past experiences which engage and integrate immediate and deliberative thinking processes. Here, Kahneman observes that expertise born out of mental practice plays an important role in honing accurate and balanced intuitive judgments and choices.

Kahneman reminds us that we have intuitive feelings and opinions about almost everything that comes our way. We like or distrust people long before we know much about them; we trust or distrust strangers without knowing why; we feel that an enterprise is bound to succeed without analyzing it.

Whether we state them or not, we often have overly confident answers to questions that we don't completely understand, relying on evidence that we can neither explain nor defend. We also fool ourselves by constructing flimsy accounts of the past failures and successes, believing they are true.[ccxxxix],[ccxl]

One system of perceptions readily indulges narrative fallacies borne out of attempts to make sense of the world through explanatory stories we find compelling and simple. As Kahneman observes, these perceptions favor pictures of reality which are concrete rather than abstract; assign larger roles to talent, stupidity and intentions rather than luck; and focus upon a few striking examples that happened, rather than on the countless events that failed to happen.

Whereas some predictive judgments rely largely on precise calculations and explicit analyses, others involve two main varieties. One form of these intuitions draws primarily upon skill and expertise acquired through repeated experience. (Chess masters and physicians are examples.) The other type, which is sometimes subjectively indistinguishable from the first, arises from oversimplifying a complex issue, such as by substituting an easier answer in response to a more difficult question that was asked.[ccxli]

Most valid intuitions develop when experts have learned to recognize familiar elements in new situations, and then to act in a manner that is appropriate. And whereas associative experience memory excels at observing patterns and relationships, it does not (cannot) allow for information it does not have. Information that is not retrieved (even unconsciously) from memory might just as well not exist.[ccxlii]

Our memories record vast repertories of knowledge and skills acquired throughout our lifetime of practice which we apply through intuition to guide us through analogous circumstances and challenges. Kahneman emphasizes that the acquisition of skills most particularly requires a regular environment, an adequate opportunity to practice, and rapid and unequivocal feedback about the correctness of thoughts and actions.[ccxliii]

Henry Simon, who studied chess masters, observed that following thousands of hours of practice they came to see pieces on the board differently from the rest of us. He wrote:

> *The situation has provided a clue; this clue has given the expert access to information stored in his memory, and the information provides the answer. Intuition is nothing more than*

recognition.[ccxliv]

Accordingly, Daniel Kahneman reminds us that whereas "expert" intuition strikes us as magical, it is not. Indeed, he writes:

> *...each of us performs feats of intuitive expertise many times each day. Most of us are pitch-perfect in detecting anger in the first word of a telephone call, recognize as we enter a room that we were the subject of the conversation, and quickly react to subtle signs that the driver of the car in the next lane is dangerous.*[ccxlv]

Thanks to our fast-thinking functions, many of these intuitive judgments are instantaneous. This is when our associative experience memory takes control to distinguish surprising from normal events in a fraction of a second, recognizes the causal nature of the abnormal surprise and automatically searches for the best response.

Associative memory involves a vast network of ideas which might be thought of as "nodes," which link together according to categories we unconsciously ascribe to them. However, instead of imagining as psychologists once did that our minds go through conscious ideas one at a time, the current view of associative memory perceives a great deal happening all at once.

While each new idea simultaneously activates many others, only a few of them register on our consciousness. As Kahneman notes, most of the work of associative thinking is silent, hidden from our conscious selves.[ccxlvi]

Thinking smart doesn't necessarily have to be hard work.

In fact, doesn't it seem that many of your best ideas come to mind when you are most relaxed and in a reflective mood?

Economist Daniel Kahneman points out that our more relaxed ideas and decisions may often be better than when we consciously work our thinking minds too hard. He observes that people who are cognitively very busy are more likely to make superficial judgments in social situations through weakened self-control. A sleepless night or a few drinks, for example, can have the same disrupting effect by overloading short-term memory.[ccxlvii],[ccxlviii]

So, if the highly generalized mental shotgun approach makes it easier to generate quick answers to difficult questions without imposing much hard work on our lazy relaxed cognitive processes, maybe that isn't always so bad. And considering our frequent overconfidence in believing things we really know very little about, the fact that we are still around to learn from these occasions suggests that none of those errors were fatal...at least not yet.[ccxlix]

In any case, if oversimplification of issues sometimes gets us into trouble, tendencies leaning too far in the other direction can as well. Kahneman writes:

> *Experts try to be clever, think outside the box, and consider complex combinations of features in making predictions. Complexity may work in the odd case, but far more often than not it reduces validity. Simple studies have shown that human decision makers are inferior to a prediction formula even when they are given the score suggested by the formula! They feel they can overrule the formula because they have additional information about the case, but they are wrong more often than not.*[ccl]

English actor Dan Stevens regards the personal "comfort zone" as the great enemy of creativity, He advises: "Moving beyond it necessitates intuition, which in turn configures new perspectives and conquers fears."

British celebrity chef, Heston Blumenthal, observes:

> *As we get older, we tend to become more risk averse because we find reasons why things won't work. When you are a kid you think everything is possible, and I think with creativity it is important to keep that naivety.*

While much creative thinking occurs subconsciously, we can often give ourselves a boost in situations that require ingenuity and flexible thinking. Sometimes this requires getting outside of our comfort zones of tried-and-true experiences, ready-made methods and popularly-assumed perspectives.

Breaking Molds and Remolding Possibilities

Novelist W. Somerset Maugham wrote that "Tradition is a guide and not a jailer." The past may be revered, but it is not untouchable. Still, all too often, education at all levels focuses excessively backwards on received knowledge and established results to the neglect of forward views toward a better world that students can design, build, and inhabit.

Psychologist/writer Edward de Bono credited with originating the term "lateral thinking" believes that "creativity involves breaking out of established patterns in order to look at things in a different way."

He counsels that "One very important aspect of motivation is the willingness to stop and look at things no one else bothered to look at. This simple process of focusing on things that are taken for granted is a powerful source of creativity."

De Bono advises:

> *We need creativity in order to break free of temporary structures that have been set up by a particular sequence of experience.*

Here, originality requires breaking some molds, including those we have allowed to crystalize around our own mindsets. Conversely, and also true, Hungarian-British author and journalist Arthur Koestler defined creativity as "breaking of habits through originality." Inventor Charles Kettering characterized innovation as to "Get off Route 35."

David Eagleman and Anthony Brandt remind us that the very origins of originality typically consciously or subconsciously draw upon raw materials provided by past experiences. Just like the massive programs running silently in computers, our inventiveness typically runs in the background, outside of our direct awareness. We consciously or unconsciously transform catalogued materials into new or modified forms by bending, breaking and blending them.[ccli]

Bending refers to remodeling a preexisting idea or object archetypes into a different version. For example, by changing its shape or size to fit another purpose. They note that this such bending takes endless forms, including chorographers who bend human forms of dancers to fit themes and movements of performances, and recording artists who create their own renditions of popular music.

Eagleman and Brandt describe breaking as taking an existing concept or thing apart and reassembling the fragments in new ways. An example is when biochemist Frederick Singer figured out how to chop complex insulin molecules into shorter, more manageable, pieces for sequencing the composite amino acids in order to figure out the overall insulin molecule architecture. His "jigsaw" method, which earned Singer a 1958 Nobel Prize, is still used to map the structure of proteins.

Blending entails combining two or more concept sources in novel ways. This often involves weaving together different threads of knowledge from the natural world to produce designs and structures that follow the same patterns and principles. As with bending and breaking, this blending can be observed in virtually all aspects of human creation and activity: in art, literature and music; in metaphors used in everyday communication; and in sciences such as chemistry and metallurgy.

Authors Eagleman and Brandt hark back to chronicle a great leap forward in human civilization which occurred around 3,300 BC when Mesopotamians first blended copper and tin to begin a revolutionary Bronze Age of weaponry, sculpture, pottery and coinage. Whereas each of the components which had been separately used thousands of years earlier are relatively soft, when mixed together, they are harder than wrought iron.[cclii]

Eagleman and Brandt conclude that we bend, break and blend everything we observe, and that those tools allow us to extrapolate far beyond the reality around us. Much of this bending, breaking and blending ideas occurs at times when we don't realize it is going on. Yet whenever conscious or not, these cognitive processes apply to all areas and practitioners of creativity that break standard molds of thought to reveal previously unseen patterns, relationships and possibilities.

American entrepreneur, business magnate, inventor and college drop-out Steve Jobs described creativity as "just connecting things." He explained:

When you ask creative people how they did

*something, they feel a little guilty because they
didn't really do it. They just saw something. It
seemed obvious to them after a while; that's
because they were able to connect experiences
they've had and synthesize new things.*

You may have observed that not all of those creative new things
we think up receive the sort of deserved enthusiastic responses
we have hoped for. Thinking "out of the box" often smacks of
unwelcome nonconformity to prevailing and popularly accepted
conventions and standards. These "new" ideas are often openly
resisted by closed mind-sets.

Social scientists Diego Gambetta and Steffen Hertog
characterize a potentially dangerous form of mind-set that
seems to be opposite of the improviser...one that is thin-
skinned and easily bruised when preconceived expectations are
not met. Ironically, they found this closed-mind condition to be
particularly common among engineering students who we
might naturally imagine to be among the most inventive
populations. The researchers even controversially suggested
that engineering programs might foster such mind-sets through
intensive emphasis on decontextualized knowledge.[ccliii]

Creativity to overcome fear and mental inertia involves
acting upon our curiosity and confidence to open our minds to
new modal possibilities.

As English screenwriter and producer John Cleese, who
co-founded the popular *Monty Python* comedy troupe,
explains:

*We all operate in two contrasting modes,
which might be called open and closed. The
open mode is more relaxed, more receptive,
more exploratory, more democratic, more*

playful and more humorous. The closed mode is the tighter, more rigid, more hierarchical, more tunnel-visioned.

Cleese observes:

Most people, unfortunately spend most of their time in the closed mode. Not that the closed mode cannot be helpful. If you are leaping a ravine, the moment of takeoff is a bad time for considering alternative strategies. When you charge the enemy machine-gun post, don't waste energy trying to see the funny side of it. Do it in the 'closed' mode.

But the moment the action is over, try to return to the 'open' mode—to open your mind again to all the feedback from our action that enables us to tell whether the action has been successful, or whether further action is needed to improve on what we have done. In other words, we must return to the open mode, because in that mode we are the most aware, most receptive, most creative, and therefore at our most intelligent.

Albert Einstein stressed that "The true sign of intelligence is not knowledge, but imagination." He also posited that:

To raise new questions, new possibilities, to regard old problems from a new angle, requires creative imagination and marks real advances in science.

Educational author George Kneller notes that such discoveries or reflective insights often appear to us at moments when we take time to reexamine our thinking to consider what circumstances or assumptions might be misconstrued. Kneller observes:

> *Creativity, as has been said, consists largely of rearranging what we know in order to find out what we do not know. Hence, to think creatively, we must be able to look afresh at what we normally take for granted.*

Hungarian-British author and journalist Arthur Koestle believed that a truly creative education comes from within. He wrote:

> *Creative activity could be described as a type of learning process where teacher and pupil are located in the same individual.*

Psychologist Stephen Nachmanovitch writes:

> *Education must tap into the close relationship between play and exploration; there must be permission to explore and express. There must be validation of the exploratory spirit, which by definition takes us out of the tried, the tested, and the homogeneous.*[ccliv]

Determination and Perseverance

English writer and poet Samuel Johnson said, "Self-confidence is the first requisite to great undertakings." This fundamentally involves believing in possibilities...overcoming doubts of failure...remaining open to new experiences...being too busy working and learning to limit boundaries of investigation and thought. It is an attitude that says: "I can do this thing...I can make this work out...this will be a worthwhile contribution and exciting learning experience."

We can readily detect confident people. They tend to be relaxed, focused and positive. Their self-assurance stems from emotional feelings about how they see and value themselves, rather than upon all-consuming concerns about how others see them. Self-confident people know themselves, recognize their own strengths and acknowledge their weaknesses as opportunities for improvement.

French philosopher, playwright, novelist, literary critic and political activist Jean-Paul Sartre warned, "If you are lonely when you're alone, you are in bad company."

How we psychologically view and value ourselves greatly influences our performance in areas of life...very much including the business world. Those who only feel good about themselves through recognition and approval by others are consigned never to be free.

Poor self-esteem virtually guarantees defeat and failure. Without it, we're unlikely to take the risk of giving everything necessary we've got to the challenge at hand. Or as nine-time Olympic gold medal winning track and field athlete Carl Lewis expressed it, "If you don't have confidence, you'll always find a way not to win."

As Albert Bandura, a psychologist at Stanford University, writes:

> *Perceived self-efficacy is defined as people's beliefs about their capabilities to produce designated levels of performance. Self-beliefs determine how people feel, think, motivate themselves and behave.*

Bandura reports that people with lots of confidence in their abilities approach difficult tasks as challenges to be mastered, rather than as threats to be avoided. They set challenging goals for themselves and maintain strong commitments to achieving them. They persevere and quickly recover after setbacks. They attribute failures to insufficient efforts, knowledge and skills that they believe are attainable.

Sometimes we judge ourselves too harshly. We get caught up in vicious cycles of declining confidence and performance. We get discouraged and engage in negative self-talk.

John Assaraf, *New York Times* bestselling author of *Having it All,* writes that while we may not always be aware of it, we constantly create and repeat self-affirmations. He explains:

> *The problem is, we typically don't pay attention to exactly what those affirmations are saying. Often we go through the day giving ourselves all sorts of contradictory, or even negative messages. We may project confidence to the world around us, while our inner dialogue says...I hope this works. I am so nervous about this. I hope I don't blow it. Affirmations are self-fulfilling prophecies. If we say, 'This is never going to work'...then chances are excellent it never will.*

Entrepreneurs recognize and accept that risks of painful setbacks and failures are very real. They realize through experience that failing can often be another step in a new and better direction. They keep moving forward and learning in the process. They look for and see a big, bright, colorful picture that others see only as a problem or obstacle.

Two businesspeople face the same tough conditions but respond to them entirely differently. One views the situation as a threat that causes stress and anxiety. The other sees the same conditions as a challenge that motivates and excites them.

Steve Jobs cautions:

> *Your time is limited, so don't waste it living someone else's life. Don't be trapped by dogma—which is living with the results of other peoples' thinking. Don't let the noise of others' opinions drown out your inner voice. And most important, have the courage to follow your heart and intuition. They somehow already know what you truly want to become. Everything else is secondary.*

The great American orator and influential politician, William Jennings Bryan, advised that "The way to develop self-confidence is to do the thing you fear and get a record of successful experiences behind you." In doing so, success validates confidence, demonstrating that one's belief in their own judgment and ability is well-founded. Success breeds success.

Marianne Williamson, a four-time *New York Times* bestselling author and founder of project Angle Food, a meals-on-wheels program serving homebound Los Angeles AIDS patients, inspires us to overcome self- doubts that handicap full lives. She writes:

> *Our deepest fear is not that we are inadequate. Our deepest fear is that we are powerful beyond measure. It is our light, not our darkness that most frightens us. We ask*

ourselves, 'Who am I to be brilliant, gorgeous, talented, fabulous?'

Actually, who are you not to be? You are a child of God. Your' playing small does not serve the world. There is nothing enlightened about shrinking so that other people won't feel insecure around you.

We are all meant to shine, as children do. We were born to make manifest the glory of God that is within us. It's not just in some of us; it's in everyone. And as we let our own light shine, we unconsciously give other people permission to do the same. As we are liberated from our own fear, our presence automatically liberates others.

Philosophy professor Stephen Asma emphasizes that an improvising mind-set readily accepts risks of failure. The true improviser gets knocked down ten times and gets up eleven.

American engineer, inventor, Delco founder and General Motors research director Charles Kettering advised that true inventors don't take either formal education or fear of failing very seriously. He wrote:

You see, from the time a person is six years old until he graduates from college he has to take three or four examinations a year. If he flunks once, he is out. But an inventor is almost always failing. He tries and fails maybe a thousand times. If he succeeds once, then he's in. These things are diametrically opposite. We often say that the biggest job we have, is to teach a newly hired employee how to fail

> *intelligently. We have to train him to experiment over and over and to keep on trying and failing until he learns what will work.*

The late American scientist and inventor Edwin H. Land, who co-founded the Polaroid Corporation, once said:

> *The essential part of creativity is not being afraid to fail.*

We can readily observe that highly recognized innovative achievements often occur following repeated unsuccessful attempts by people with tendencies to obstinately persevere in a course of action in spite of difficulty or opposition.

Consider, for example, that day in 1879, when Thomas Edison applied electricity to a fine thread of carbon thread that he twisted into a horseshoe shape. Although the filament glowed steady and bright, he recognized that it was inadequate to market as a commercially viable bulb.

"Ransacking nature's warehouse," Edison set out to experiment with alternative filament materials including various plants, pulp, cellulose, flour paste, tissue paper and synthetic cellulose.[cclv]

Edison later said:

> *I speak without exaggeration when I say that I have constructed 3,000 different theories in connection with the electric light, each one of them reasonable and apparently likely to be true. Yet only in two cases did my experiments prove the truth of my theory.*

Not all of Edison's ideas panned out that great...He wanted to offer the middle-class public a much more affordable piano than a Steinway. The one he designed in the 1930s out of concrete for the Lauter Piano Company failed miserably. Weighing literally a ton, its sound quality was also far inferior.[cclvi]

Some idea seedlings just require a lot of patient nurturing before they bear fruit. James Dyson's invention of the first successful bag-less vacuum cleaner required 5,127 prototypes and fifteen years before going to market. He reported:

> There are countless times an inventor can give up on an idea. By the time I made my fifteenth prototype, my third child was born. By 2,627, my wife and I were really counting our pennies. By 3,727, my wife was giving art lessons for some extra cash. These were tough times, but each failure brought me closer to solving the problem.[cclvii]

Or as Samuel Beckett advised, "Try again. Fall Again, Fall Better."

David Eagleman and Anthony Brandt remind us that failed ideas aren't really wasted. Many of them serve as fodder for other concepts and creations:

> Innovation takes wing when the brain generates not just one new scheme, but many, and sketches those ideas to different distances from what is already known and accepted. Risk-taking and fearlessness in the face of error propel those imaginative thoughts.[cclviii]

American author Roger von Oech, whose focus has been on the

study of creativity, observes:

> *It's easy to come up with new ideas; the hard part is letting go of what worked for you two years ago, but will soon be out of date.*[cclix]

Linus Pauling, one of the founders of the fields of quantum chemistry and molecular biology, reportedly advised:

> *The best way to have a good idea is to have a lot of ideas.*[cclx]

And as British molecular biologist, biophysicist and neuroscientist Francis Hampton Compton Crick reportedly once said:

> *The dangerous man is the one with only one theory, because he'll fight to the death for it...the stronger approach is to have lots of ideas and let most of them die.*[cclxi]

We can observe that successful people around us who both recognize and act upon opportunities that most other people let pass by. Here, former Harvard Business School professor and *Harvard Business Review* editor Theodore Levitt offers an important distinction between creativity and innovation, whereby:

> *Creativity is thinking up new things. Innovation is doing new things.*

Countless well-known people have earned their lofty achievement status "the hard way:" through extraordinary effort and perseverance. In doing so, many have exhibited heroic persistence and resilience in repeatedly overcoming daunting obstacles. Some have rallied above poverty and financial setbacks, physical and learning challenges, mental depression, ideological peer resistance and social discrimination and even life-threatening religious and political strife.

We might assume that virtually all have known repeated failures: concepts that didn't pan out as hoped, theorems that couldn't be validated and ideas that were ahead of their time. And while some experienced recognition and rewards during their lifetimes, others who weren't nearly so fortunate bequeathed the richness of their legacies to us.

Calvin Coolidge, the 30th President of the United States, a man noted for decisive actions, attributed central importance to dogged determination. He said:

> *Nothing in this world can take the place of persistence. Talent will not: nothing is more common than the unsuccessful men with talent. Genius will not: unrewarded genius is almost a proverb. Education will not: the world is full of educated derelicts. Persistence and determination alone are omnipotent.*

Albert Einstein reportedly said:

> *It's not that I'm so smart, it's just that I stay with problems longer.*

Michelangelo is also credited with modestly describing his genius as more a matter of eternal perseverance and patience, stating:

If people knew how hard I worked to get my mastery, it wouldn't seem so wonderful.

Regarding the importance of diligence and discipline, Nikola Tesla wrote:

> *I am credited with being one of the hardest workers and perhaps I am, if thought is the equivalent of labor, for I have devoted to it almost all of my waking hours. But if work is interpreted to be a definite performance in a specified time according to a rigid rule, then I may be the worst of idlers. Every effort under compulsion demands a sacrifice of life energy. I have never paid such a price. On the contrary, I have thrived on my thoughts.*[cclxii]

The *Merriam-Webster* dictionary defines persistence as "the quality that allows someone to continue doing something or trying to do something even though it is difficult or opposed by other people." In other words, it typically requires powerfully determined commitment, often under outright discouraging circumstances.

As characterized by America's 32nd President Franklin Delano Roosevelt:

> *When you reach the end of your rope, tie a knot in it and hang on.*

Examples of successful perseverance are endless. J.K. Rowling lived and raised her young daughter under government assistance in the United Kingdom while writing her first *Harry Potter* book, which was rejected by 12 major publishing houses over a seven-year period.

Henry Ford, who went bankrupt three times before he produced his first successful automobile, said:

> *Failure is merely an opportunity to more intelligently begin again.*

Winston Churchill observed that: "Success consists of going from failure to failure without loss of enthusiasm." He emphasized: "Continuous effort—not strength or intelligence—is the key to unlocking our potential."

Famed American architect, systems theorist and inventor Buckminster ("Bucky") Fuller, said:

> *I'm not a genius. I'm just a tremendous bundle of experience...Most of my advances are by mistake. You uncover what is when you get rid of what isn't." He also reflected: "How often I found where I should be going only by setting out for somewhere else.*

Thomas Edison, as quoted in a 1921 interview by B.C. Forbes for *American Magazine*, said:

> *I never allow myself to become discouraged under any circumstances. I recall that after we had conducted thousands of experiments on a certain project without solving the problem, one of my associates after we had conducted the crowning experiment and it had proved a failure, expressed discouragement and disgust over our having failed to find out anything. I cheerily assured him that we had learned*

something. For we had learned for a certainty that the thing couldn't be done that way. And we would have to try some other way. We sometimes learn a lot from our failures if we have put into the effort the best thought and work we are capable of.

Professional basketball star, performer, businessman and chairman of the Charlotte Hornets Michael Jordan cautions that sometimes we're inclined to expect winning results too soon. He has said:

I've failed over and over again in my life...that is why I succeed.

The idea, of course, is to learn from mistakes...what went wrong...what parts succeeded...rather than perpetuate them.

South-African-American business magnate, inventor and SpaceX founder-CEO Elon Musk stresses the importance of constant performance reassessments and adjustments in personal and corporate life:

I think it's very important to have a feedback loop, where you're constantly thinking about what you've done and how you could be doing it better. I think that's the single best piece of advice: constantly think about how you could be doing things better and questioning yourself.

Douglas Hofstadter, Pulitzer Prize-winning author for cognitive sciences ("Hofstadter's Law") plus composer, artist, calligrapher, physicist and programmer, summarizes the general process:

> *You make decisions, take actions, affect the*
> *world, receive feedback from the world,*
> *incorporate it into yourself, then the updated*
> *'you' makes more decisions, and so forth,*
> *round and round.*

W. Arthur ("Skip") Porter, a Texas business executive who served as Oklahoma Secretary of Science and Technology, characterized "The innovation point [as] the pivotal point when talented and motivated people seek the opportunity to act on their ideas and dreams."

John Mighton, who barely passed calculus in a freshman-level university course and later earned a Ph.D. in mathematics and advanced the pioneering knot and graph theory, and also an award-winning playwright, attributes his achievements to breaking a task into a series of steps and then practicing repeatedly. He observes:

> *People with expert abilities are generally*
> *made, not born and often their abilities arise*
> *out of a great deal of repetitive practice and*
> *imitation and copying of other peoples' styles*
> *and ideas. For instance, chess masters*
> *repeatedly play small sets of moves, memorize*
> *thousands of positions and obsessively study*
> *the games if the masters.*[cclxiii]

In his book *Talent is Overrated: What Really Separates World-Class Performers from Everybody Else*, Geoff Colvin also emphasizes the fundamental importance of repeated practice to success...the sort of practice that we don't necessarily regard to be unpleasant when we're doing it. With motivation and discipline, we can become much better at anything we do.[cclxiv]

Colvin, a senior editor-at-large for *Fortune* magazine, acknowledges that certain inherited traits influence advantages or disadvantages in some fields. In sports, for example, "a five-footer will never be an NFL lineman, and a seven-footer will never be an Olympic gymnast."

Yet Colvin also points out that we are not generally held hostage to some naturally granted level of talent:

> *We can make ourselves what we will. Strangely, that idea is not popular. People hate abandoning the notion that they would coast to fame and riches if they found their talent. But that view is tragically constraining, because when they hit life's inevitable bumps in the road, they conclude that they just aren't gifted and give up.*

Why don't people shoot for the stars? Colvin answers, "Because for most people, work is hard enough without pushing even harder. Those extra steps are so difficult and painful they almost never get done."

Geoff Colvin recognizes that "maybe we can't expect most people to achieve greatness because it's just too demanding. That's the way it must be. If great performance were easy, it wouldn't be rare."

As to why some people are motivated to take those extra steps while most aren't, Colvin leaves this question unanswered. Yet he offers great hope for all who have discovered and acted upon that passion to excel:

> *...the striking, liberating news is that greatness*

isn't reserved for a preordained few. It is available to you and to everyone." He concludes that while "Talent has little or nothing to do with greatness. You can make yourself into any number of things, and you can even make yourself great.[cclxv]

Yes...but we each might ask, great at what?

Maybe it can be greater at making our personal lives more satisfying and fulfilling for ourselves and others. Or perhaps we can become greater at doing whatever we're doing for whatever reasons we are choosing to do it...or possibly making more time for other opportunities and priorities to do even more.

Maybe by recognizing greatness in examples of others we can rediscover exceptional potentials in ourselves that motivate us to set greater goals that test our limits...and to perhaps challenge our doubts whether any such limits actually exist.

And maybe we can discover that we are already pretty great and simply want to get better at it.

REINVENTING OURSELVES

SO, WHAT ABOUT getting better at thinking by inventing artificial brains that are smarter than we are to teach us, to expand our limited biological faculties ...or maybe even replace our soon-to-be obsolete, inferior human organic models?

And after all, what characteristics will characterize "real" intelligence versus the evolution of artificial versions? Will computers always continue to depend upon information systems and data we teach them?

Might rapid computing evolutions ultimately revolt against us HAL did against the Discovery One astronaut crew in the movie *2001: A Space Odyssey* based upon Arthur C. Clarke's novel series?

As you may recall, HAL 9000 (the Heuristically programmed Algorithmic computer) decided to kill the astronauts by locking them outside the spacecraft during a repair after reading their lips to discover a secret plan to disconnect the robotic system's cognitive circuits following lack of trust triggered by a computing glitch. HAL also attempts to suffocate hibernating onboard crewmembers by disconnecting their life support systems.

Fortunately, in this case, human dexterity wins the day. One of the astronauts circumvents HAL's control by manually opening an emergency airlock, detaching a door via its explosive bolts, reentering Discovery, and quickly re-pressurizing the airlock.

Whew!

Nevertheless, can we truly blame HAL for lack of empathy or gratitude for its mentally inferior human creators? If they were smarter, wouldn't HAL's programmers recognize the risk that artificial intelligence might inevitably surpass capabilities of our human "general intelligence"?

Far-fetched?

Actually, some computer algorithms already teach themselves, and in the process, learn much faster than we do.

Thinking Machines?

A great deal of progress has occurred since the concept of "thinking machines" was first hypothesized during a 1956 meeting of scientists, mathematicians and engineers at Dartmouth College is no longer theoretical.

Moreover, those machines are already beginning to outthink some top human experts in certain very complicated mental challenges.

- By 1997, a "Deep Blue" IBM computer defeated the reigning world chess champion, Gary Kasparov.
- In 2011, "Watson," another IBM computer, beat all humans in the quiz show *Jeopardy*.
- In 2016, an "AlphaGo" algorithm developed by "DeepMind," a London AI company, dispatched Lee Sedol, a top player in ancient and complex board game

"Go." The algorithm was originally trained on 160,000 games from a database of previously played games. The program was later upgraded to "AlphaGo Zero," which taught itself by playing four million games against itself entirely by trial and error. AlphaGo Zero subsequently annihilated its parent, AlphaGo, 100 games to zero. It accomplished learning capacity in less than one month that would have required a decade or two of training for a human to become a highly skilled Go master.

- In 2017, "Libratus" software developed at Carnegie Mellon University beat four top players over a 20-day tournament of No-Limit Texas Hold'em poker. The code doesn't need to bluff...it just outthinks humans.[cclxvi]

We have witnessed a worldwide impact of artificial intelligence over just the last few years to the point that it dominates nearly all businesses, investments and even ethical narratives. As a result, two opposing attitudes appear to have emerged: one believing that AI will beneficially augment humans and the other that it will diminish them. Most likely prospects hold that both predictions are true.

There is virtually no likelihood that the AI revolutionary march of encroachment upon the human domain of activities will lose momentum. As Professor Justin Zobel, head of the department of Computing & Information Systems at the University of Melbourne, Australia observes:

> *It is a truism that computing continues to change our world. It shapes how objects are designed, what information we receive, how and where we work, and who we meet and do business with. And computing changes our*

understanding of the world around us and the universe beyond.[cclxvii]

Zobel notes, for example, that while computers were initially used in weather forecasting as no more than an efficient way to assemble observations and do calculations, today our understanding of weather is almost entirely mediated by computational models.

Another example relates to sweeping influences upon biological sciences and commercialization. Zobel points out that whereas research was once done entirely in the lab (or in the wild) and then captured in a model, it often now begins in a predictive model, which then determines what might be explored in the real world.

Justin Zobel alerts us to a new AI reality that computers are not only influencing a reinvention of ourselves, but also rapidly evolving towards a capacity to reinvent themselves. Here, he urges us to recognize that computing influences upon human activities and society often characterized as "digital disruption" are also disrupting the very nature and future of digital computing.

Rather than to perceive AI as a singular revolution, it is more appropriately conceptualized as an exponentially expanding series of independently-perpetuating evolutions.

Professor Zobel directs our attention to a phenomenon whereby AI evolution is simultaneously advancing on separate, diverse and revolutionary tracks. He writes:

> *Each wave of new computational technology has tended to lead to new kinds of systems, new ways of creating tools, new forms of data, and so on, which have often overturned their predecessors. What has seemed to be*

270

evolution is, in some ways, a series of revolutions.[cclxviii]

Zobel emphasizes that this technological explosion of computing computational capacities and applications is more than a single chain process of innovation which has traditionally been a hallmark of other physical technologies that shape our world. He cites as a previous example, a chain of inspiration from waterwheel, to steam engine, to the internal combustion engine.

Underlying all of this was a process of enablement...the industry of steam engine construction yielded the skills, materials and tools used in construction of the first internal combustion engines.

A big and consequential difference here is that in computing, something richer is happening where new technologies emerge...often not only replacing their predecessors, but also enveloping them. In doing so, computing is creating platforms on which it reinvents itself, continuously reaching up to the next platform.

Imagine how rapidly this evolutionary history has advanced in a mere half-century since the so-called "digital computer revolution" first emerged in the commonplace public lexicon to describe the new phenomenon of rapid micro-processing during the 1970s and 1980s!

First, the invention of integrated circuit (microchip) technology greatly reduced the size and cost of their enormously expensive mainframe predecessors. Then, after "chip-on-a-chip" was commercialized, the cost to manufacture a computer system again dropped dramatically.

Arithmetic, logic and control functions that had previously occupied several costly circuit boards became available in a single integrated circuit which made high-volume manufacture possible. Concurrently, solid state memory advancements eliminated bulky, costly and power-hungry magnetic core systems used in previous generations.

Early advancements were generally created by independent entities and led to the availability of cheap and fast computing, affordable disk storage and networking. Within only a decade, computers became common consumer goods for word processing and gaming. Computing and information storage were contained in personal standalone units.

Speaking at a 1977 World Future Society meeting, Digital Equipment Corporation CEO Ken Olsen famously said, "There is no reason for any individual to have a computer in his home." [cclxix]

Olsen couldn't have been more wrong.

By the late 1980s, personal computers increasingly earned their place in private homes and businesses. Families, for example, found "kitchen computers" convenient for storing easily retrievable disk-based recipe catalogs, medical databases for childcare, financial records and encyclopedias for schoolwork.

Although predicted to be commonplace before the end of the decade, computers still weren't powerful enough to match more optimistic visions. Due to limited memory capacities they could not yet multitask, floppy disk-based storage was inadequate both in capacity and speed for multimedia and display graphics were blocky and blurry with jagged text.

Networking technologies created in university computer sciences departments soon led to substantial collaborative software improvements. The resulting emergence of an open-source information culture then spread throughout wide user communities which took advantage of—and also contributed to—common operating systems, programming languages and tools.

It took only another decade for computers to mature sufficiently for graphical user interfaces to serve broad, non-technical user markets which gave rise to the Internet. Equipment and user costs dropped dramatically as data catalogs became maintained online and accessed over the World Wide Web rather than stored on floppy disks or CD ROM.

The global digital traffic infrastructure Internet formed as networks became increasingly more uniform and interlinked. Simultaneous increases in computing power and falling data storage costs rapidly expanded world-wide service markets. The Internet, which was popularized for personal email chat and business forums, also became a growing exchange mechanism for computer data and codes.

The marvelous confluence of networking, capacity and storage which began in the 1990s in combination with an open-source culture of sharing both leading and drawing from the Internet remains in the infancy of a yet unknown evolutionary creature. It is highly speculative bordering on pure fantasy to imagine what forms will emerge years, much less decades in the future.

On the basis of computer capacity alone, American engineer and Intel co-founder Gordon Moore made a prediction in 1965 that the number of transistors per silicon chip has doubled every year and would continue to do so over the following decade. By his estimation at that time, microcircuits of 1975 would contain an astounding 65,000 components per chip.

By 1975, as the rate of growth began to slow, Moore revised his time frame to two years. This time his revised law was a bit pessimistic; over roughly 50 years from 1961, the number of transistors had doubled approximately every 18 months. Nevertheless, Moore's estimates were broadly accorded great importance as virtual law.

Moore had based his prediction upon a dramatic explosion in circuit complexity made possible by steadily shrinking sizes of transistors over the decades. Measured in millimeters in the late 1940s, the dimensions of a typical transistor in the early 2010s were more commonly expressed in tens of nanometers (a nanometer being one-billionth of a meter)—a reduction factor of over 100,000.

Transistor features measuring less than a micron (a micrometer, or one-millionth of a meter) were attained during the 1980s, when dynamic random-access memory (DRAM) chips began offering megabyte storage capacities.

At the dawn of the 21st century, these features approached 0.1 micron across, which allowed the manufacture of gigabyte memory chips and microprocessors that operate at gigahertz frequencies. Moore's law continued into the second decade of the 21st century with the introduction of three-dimensional transistors that were tens of nanometers in size.

Problems with Moore's original capacity estimate which surfaced in 1975 had encountered a technical snag due to limitations posed by the photolithography process used to transfer the chip patterns to the silicon wafers which used light with a 193-nanometer wavelength to create chips with features just 14 nanometers.

A roadmap around chip limitations sometimes described as "More than Moore" that followed applies highly integrated chips which combine a diverse array of sensors and low-power processors. This led to the growth of smartphones and the "Internet of Things". These processors include RAM, power regulation, analog capabilities essential for GPS, cellular telephones and Wi-Fi radios.

Something else to expect...Moore's innovative lawbreakers will continue to produce computing processors which are more versatile, faster, smaller and energy-efficient in response to ever-growing demands for smarter systems that support and compete with human enterprises.

In May 2019, OpenAI, a coalition of 100 tech experts in San Francisco whose mission is "discovering and enacting the path to safe artificial general intelligence," reported that the power of machine learning operations has doubled about every 14 weeks. That's a blistering pace compared to Moore's Law.

A technique called "Generative Adversarial Networks" (GANs) trains competing AI algorithms to challenge each other, learn from mistakes and even to fool one another with convincing deceptions.[cclxx]

GANs is an example within a broad AI technology field broadly referred to as machine learning which essentially mimics the way we learn from trial and error using positive and negative reinforcement methods to achieve a desired outcome. The process uses two opposing reward or loss functions; one is a generative model (also known as the environment), and the other is a discriminative model (also known as an agent).

In one example, a GANs generator randomly creates images that the discriminator must identify as either real or recognize to be an artificial fake. Both entities are trained over a large number of iterations, with each iteration improving the "skill" ability of each. Over time, the discriminator learns to ever-more reliably tell fake images from real images, while the generator uses the feedback from the discriminator to learn to produce more convincing fake images.

Another example of a multi-agent GAN is "Style Transfer" where the model is provided two photos and two discriminators are tasked to produce a single picture. One of the discriminators is rewarded by conserving the content of the first image, while the other is rewarded by preserving the style of the second image. In the case of one discriminator presenting an abstract pattern—and the other presenting a picture of an elephant—each will work hard to come up with a compromise which satisfies both. Applying loss/reward criteria, a single picture will result after perhaps millions of iterations.

Artificial intelligence will likely witness enormous leaps forward with the advent of staggeringly large computing capacities afforded by rapidly accelerating quantum computer (QC) advancements.

As University of Maryland researcher Christopher Moore testified at an October 2017 House Science Committee hearing on "American Leadership in Quantum Technology," merely 300 atoms under full quantum computer control might potentially store more pieces of information than the number of atoms that exist in the entire Universe.[cclxxi]

Such implausible features are made possible by equally incomprehensible subatomic-scale phenomena. Unlike current computers which process tiny "bits" of data in a linear sequence as either a one or a zero, at the seemingly weird subatomic scale, a quantum bit (or "qubit") can be both a zero and a one at the same time. As a result, rather than growing linearly, adding more qubits expands computing power exponentially.

Quantum systems are potentially capable of computational feats that have proven to be inconceivable with conventional technologies. For example, they might be used to model molecules which are ridiculously hard to model with a classical computer...including trying to simulate the behavior of the electrons in even a relatively simple molecule, which is enormously complex. IBM researchers used a quantum computer with seven qubits to model a small molecule made of three atoms.[cclxxii]

IBM has a long history of quantum computer research and believes that although QC is still in early days of development, it is one of the future's most promising technologies. While no commercial grade computer has yet been built, IBM has reported that more than 100 organizations are now using its quantum computing services, including businesses, universities and government research facilities. Network clients pay to use some of the company's 15 early-stage machines via the cloud.

And while quantum computers are potentially much more powerful than traditional computers, they are also more delicate and prone to faults.[cclxxiii]

Researchers at Alphabet Inc.'s Google declared in 2019 that they had made a "quantum supremacy" breakthrough—the ability to solve a problem with a quantum computer that a regular machine couldn't muster in a reasonable time. Google's quantum computer, at a lab near Santa Barbara, Calif., had performed a mathematical operation in 3 minutes and 20 seconds that would have taken a supercomputer more than 10,000 years to complete.[cclxxiv]

Despite some progress in demonstrating quantum computing's potential, many of the calculations now being performed in labs can still be done much faster with traditional computers. Jim Clarke, director of quantum hardware at Intel Labs, Intel's research arm, told the *Wall Street Journal* that it could take many years to produce a quantum computer that is better than today's digital counterparts.

Just as AI promises to transform an endless variety of peaceful information and problem-solving tasks, its vast capacities to out-think conventional computers present enormously troubling cybersecurity challenges. In addition to overwhelming cryptographic codes used to protect top secret data, it can also be weaponized for conduct of armed conflicts at much larger scales and higher speeds than humans can comprehend or react to.

International foes and friends are racing to achieve QC supremacy which can defeat all current-generation defenses against military, information security, banking and utility infrastructure system cyberattacks. The first hostile nation to win this race will be able to open the encrypted secrets of every country, company and person on the planet; dominate global information-technology and the global financial systems; compromise the safety of medical, food and water services; put transportation and energy infrastructures at risk; and threaten domestic and military security systems.

Researchers Daniel Bernstein at the Technische Universiteit Eindhoven and Tanja Lange at the University of Illinois, Chicago stressed this urgency in a September 2017 report in the scientific journal *Nature*, noting:

> *We are in a race against time to deploy post-quantum cryptography before quantum computers arrive.*

They predict that:

> *Many commonly used cryptosystems will be completely broken once large quantum computers exist.*[cclxxv]

Writing in the *Wall Street Journal*, Committee for Justice President Curt Levy and Ryan Hagemann at the Niskanen Center posit a challenge in ensuring ways to ensure that future AI algorithms with minds of their own remain accountable to transparent oversight.

The authors' greatest concern isn't that advanced computers we create will go rogue and turn against us like HAL in *2001: A Space Odyssey*. They foresee a greater threat that AI complexity enables developers to secretly "rig" a system to the advantage of special interests, such as to manipulate a corporate operating program to reveal trade secrets to outside competitors.[cclxxvi]

The Information Revolution

Powered by Artificial Intelligence (AI), the Internet of Things, and more recently, the emergence of quantum computing, today's society is now experiencing the earliest beginnings of an inevitable, irreversible and unfathomably impactful information revolution.

As discussed in my book *Reinventing Ourselves: How Technology is Rapidly and Radically Transforming Humanity,* enthusiastic AI proponents promise tantalizingly utopian visions previously conceivable only in the fertile imaginations of fiction writers but decades, or even a few years ago.

Optimistic "experts" argue that just like previous technological revolutions, this one will again create new types of rewarding jobs, careers and lifestyle choices supported by unbounded information access and social networking resources.[cclxxvii]

Others anticipate each of these technological enticements serving more as one-way pathways to disruptive unemployment and income inequality impacts along with ever-increasing personal privacy sacrifices leading to ant farm societies.

Austrian economist Joseph Schumpeter coined the term "creative destruction" to characterize the way technological progress in the late 1940's improved the lives of many, but inevitably, only at the expense of a smaller few. As improvements to manufacturing processes such as assembly lines benefited the general economy and overall individual lifestyles, craft and artisan producers were displaced.

Writing in *Investopedia.com,* Economic sociologist Adam Hayes optimistically posits that while some industries and work roles will indeed fall as casualties of new technologies, they will be replaced by even greater, more open-ended opportunities.

Matthew Randall, the executive director of York College's Center for Professional Excellence, writes in TechCrunch.com that the trend for industrial robots replacing human manufacturing jobs is ultimately a good thing:

> In the last century, we moved from people manually building cars to robots assembling cars. As a result, manufacturers both produce more cars and employ more people per car than before. Instead of performing dangerous tasks, those workers now program the robots to do the dirty work for them—and get paid more for doing so. as long as we've had technology, we've had Luddites who literally destroy technological advancements—and yet, here we are, more productive, with higher quality of living than ever.

Randall argues:

> [In] reality, [robots] will enable us to keep more (and better) jobs at home, to grow our local industry, to improve our lives at the micro and macro levels. With greater automation, efficiency, safety and productivity, the North American manufacturing sector will not only survive, it will showcase the power of our innovation and ingenuity.

He concludes:

> So, will a robot take your job? Maybe. But in

> *return, you—and your children and*
> *grandchildren—will likely find more*
> *meaningful work, for better pay. Sounds like a*
> *good trade-off to me.*[cclxxviii]

Researchers at McKinsey & Company, a leading business consulting firm, project that a large need for employment will continue for human cognitive abilities. They conclude in a 2017 report titled *Jobs lost, jobs gained: What the future of work will mean for jobs, skills and wages*:

> *Workers of the future will spend more time*
> *on activities that machines are less capable of,*
> *such as managing people, applying expertise,*
> *and communicating with others. They will*
> *spend less time on predictable physical*
> *activities and on collecting and processing*
> *data, where machines already exceed human*
> *performance. The skills and capabilities*
> *required will also shift, requiring more social*
> *and emotional skills and more advanced*
> *cognitive capabilities, such as logical reasoning*
> *and creativity.*[cclxxix]

The McKinsey & Company researchers suggest that as a greater percentage of populations live longer, significantly larger new demands will result for a range of health care occupations, including doctors, nurses, health technicians, nursing assistants and personal home-care aids.

New careers and jobs will also be created in technology development and information technology services. While this will be a relatively small number compared to employment in healthcare and construction, they will more typically be higher-wage occupations.

According to McKinsey forecasts, broader earning challenges lie ahead from a wider societal perspective. Although some demands for lower wage occupations will increase, a wide range of middle-income occupations will suffer the largest employment declines. As a result, income polarization may continue to expand.[cclxxx]

How many jobs will be lost to AI-related technologies?

Researchers for the *MIT Technology Review* who surveyed projections by various groups regarding job losses (and some gains) at the hands of AI, automation and robots and couldn't find any consensus. They concluded:

> *There are as many opinions as there are experts...prognostications provided by companies, think tanks and research institutions are all over the map.*
>
> *Predictions range from optimistic to devastating, differing by tens of millions of jobs even when comparing similar time frames. Many focused on losses in one industry...or results of a single technology such as autonomous vehicles.*
>
> *There is only one meaningful conclusion: we have no idea how many jobs will actually be lost to the march of technological progress.*[cclxxxi]

Rewiring Societal Networks and Values

There is no turning back the clock or holding back the advances on the myriad of ways that information technology—AI and Internet connectivity in particular—are changing not only our work roles and lifestyles, but also our fundamental perceptions of who we are and our relationships to social communities.

Manuel Castellas, professor and chair of Communication Technology at the University of Southern California, observes that these impacts are not entirely either good or bad:

> *People, companies and institutions feel the depth of this [wireless communication] technological change, but the speed and scope of the transformation has triggered all manner of utopian and dystopian perceptions that, when examined closely through methodologically rigorous empirical research, turn out not to be accurate.*[cclxxxii]

Posting in *Technology Review*, Castellas observes that our current "network society" is a product of the digital revolution that has led to major sociocultural changes. One of these is the rise of the "me-centered society" which has both positive and negative aspects. This shift is marked by an increased focus on individual growth, but with an attendant decline in "community" in terms of space, work, family and ascription in general.

As Professor Castellas explains:

> *Today, social networking sites are preferred platforms for all kinds of activities, both business and personal, and sociability has dramatically increased—but it is a different kind of sociability. Most Facebook users visit the site daily, and they connect on multiple dimensions, but only on dimensions they choose. The virtual life is becoming more social than the physical life, but it is less a*

virtual reality than a real virtuality, facilitating
real-life work and urban being.

In his book *Sapiens: A Brief History of Humankind,* historian
author Yuval Noah Harari points out that a truly exceptional
human trait is realized through networking communication to
create to create new realities out of imagined fictions. All other
animals are limited to using their communication systems to
describe realities.

Harari wrote in an exchange of emails with *Smithsonian
Magazine* senior editor Arik Gabbi:

> *The Sapiens secret of success is large-scale*
> *flexible cooperation. This has made us masters*
> *of the world. But at the same time it has made*
> *us dependent for our very survival on vast*
> *networks of cooperation. This process has*
> *accelerated over the millennia, so that today*
> *nearly all the things we need for survival are*
> *provided by complete strangers.*
> *I don't know how to produce the food that I*
> *eat, how to sew the clothes I wear, or how to*
> *build the house in which I live. I write history*
> *books, get paid for it, and buy 99 percent of*
> *what I need from strangers. It is no wonder*
> *that the size of the Sapiens brain has been*
> *decreasing over the last 10,000 years.*[cclxxxiii]

Yuval Noah Harari proposes that large-scale connections
through AI and the Internet are analogous to framing machine
ethics as the basis of a new religion.

He writes:

No matter whether they believe in divine laws or natural laws, all religions have exactly the same function: to give stability to human institutions. Without some kind of religion, it is simply impossible to maintain social order.

> *During the modern era religions that believe in divine laws went into eclipse. But religions that believe in natural laws became ever more powerful. In the future, they are likely to become more powerful yet. Silicon Valley, for example, is today a hot-house of new techno-religions, which promise us paradise on Earth with the help of new technologies. From a religious perspective, Silicon Valley is the most interesting place in the world.*

AI advancements have led us to thresholds of strange new worlds of contemplation beyond familiar references of human experience. Quantum theory, for example, has uprooted traditional Newtonian views of a Universe where time is linear, gravity "pulls," space has measurable dimensions, or even that a singular "reality" exists outside the influence of our individual thoughts.

As postulated in the *Stanford Encyclopedia of Philosophy*, information technology either constitutes or is closely correlated with what constitutes our existence and the existence of everything around us, including the manner in which the Universe operates. This realization has given rise to the new fields of Information Philosophy and Information Ethics.[cclxxxiv]

Transformational innovations of this revolutionary information era are applying observed, yet poorly understood, principles to create thinking machines with seemingly limitless capacities. Such inventions are already extending—even redefining—the meaning of "artificial intelligence."

Despite skeptical views Stephen Hawking and Albert Einstein have each expressed about God and religion, both have delved deeply into mysterious workings of nature at a subatomic level which, to our conventional senses, take on extrasensory, supernatural manifestations.

The "new science" of quantum mechanics goes so far as to suggest a "preposterous" possibility that everything in the physical Universe exists only as illusory inventions of our individual minds. Whereas this concept presents a radical departure from traditional Western thought, it doesn't seem nearly so alien to much older Eastern philosophies.

Generally speaking, whereas Western philosophies tend to emphasize learning new things about what reality is, ancient Hindu and Buddhist literature speaks of removing veils of ignorance that stand between us and what we really are. And where Western religions tend to envision a Universe divided into separate material and spiritual aspects, Eastern teachings make no dichotomous distinctions between material and spiritual manifestations.

Quantum mechanics challenges any notion of material reality altogether, making no distinction between mass (quanta) and their energetic and mysteriously unpredictable relationships with individual observers. In doing so, it has yielded replicable evidence that powers of mind over matter, and realities much stranger than presumed fictions, can no longer be casually dismissed merely as quack clichés.

So, let's imagine some possible ethical and moral dilemmas as "beyond material reality" hyper-intelligent systems begin to exert more and more influence over humanity.

The late Nobel laureate Stephen Hawking warned about a desperate race to ensure that humans remain masters, not slaves, of tomorrow's supercomputers:

> *Success in creating AI would be the biggest event in human history. Unfortunately, it might also be the last—unless we learn how to avoid the risks.*
>
> *Used as a toolkit, AI can augment our existing intelligence to open up advances in every area of science and society. However, it will also bring dangers. While primitive forms of artificial intelligence developed so far have proved very useful, I fear the consequences of creating something that can match or surpass humans.*
>
> *The concern is that AI would take off on its own and redesign itself at an ever-increasing rate. Humans, who are limited by slow biological evolution, couldn't compete and would be superseded. And in the future, AI could develop a will of its own, a will that is in conflict with ours.*
>
> *Others believe that humans can command the rate of technology for a decently long time, and that the potential of AI will be realized. Although I am well known as an optimist regarding the human race, I am not so sure.*[cclxxxv]

Yuval Noah Harari shares Hawking's concerns:

> *Given current technological advances, it seems unavoidable that humans will disappear in a century or two. I don't think we will be*

destroyed in some nuclear or ecological catastrophe. Rather, I think that we will upgrade ourselves into something completely different. Humans are going to acquire abilities that were traditionally thought to be divine abilities.

Humans may soon be able to live indefinitely, to design and create living beings at will, to surf artificial realities with their minds according to their wishes. The most amazing thing about the future won't be the spaceships but the beings flying them. This will result in enormous new opportunities as well as frightful new dangers.

There is no point being optimistic or pessimistic about it. We need to understand that this is really happening—it is science rather than science fiction—and it is high time we start thinking about this very seriously.[cclxxxvi]

Documentary filmmaker and author James Barrot sounded alarm about the future in the title of his book *Our Final Invention: Artificial Intelligence and the End of the Human Era.*[cclxxxvii]

Eric Hendry explored the reasoning behind Barrot's ultimate human doom scenario in an interview he posted in the Smithsonian titled *What Happens When Artificial Intelligence Turns On Us?"*

Barrot begins by reminding us that AI is a dual-use technology and like nuclear fission, it is capable of great good or great harm.

He warns that we're just starting to see the harm:

*The [US National Security Agency] NSA used
its power to plumb the metadata of millions of
phone calls and the entirety of the internet—
critically, all email. Seduced by the power of
data-mining AI, an agency entrusted to protect
the Constitution instead abused it. They
developed tools too powerful for them to use
responsibly.*

James Barrot then highlights another current ethical battle
regarding development and deployment of advanced fully
autonomous AI-powered killer drones and battlefield robots
that don't have humans in the decision loops. He opines:

*[The policy conflict is] brewing between the
Department of Defense and the drone and
robot makers who are paid by DoD, and the
people who think it's foolhardy and immoral to
create intelligent killing machines.*

*Those in favor of autonomous drones and
battlefield robots argue that they'll be more
moral—that is, less emotional, will target
better and be more disciplined than human
operators. Those against taking humans out of
the loop are looking at drones' miserable
history of killing civilians, and involvement in
extralegal assassinations. Who shoulders the
moral culpability when a robot kills? The robot
makers, the robot users, or no one?*

Barrot speculates that in the longer term, AI approaching
human-level intelligence will neither be easily controlled or
possess a benevolent nature towards its creators. Quoting AI

theorist Eliezer Yudkowsky of MIRI [the Machine Intelligence Institute]:

> *The AI does not love you, nor does it hate you, but you are made of atoms it can use for something else.*

Barrot concludes:

> *If ethics can't be built into a machine, then we'll be creating super-intelligent psychopaths, creatures without moral compasses, and we won't be their masters for long.*[cclxxxviii]

Some commentators have suggested that the solution is to embed human values deeply inside the electronic DNA of AI machines. It's not a bad idea until you consider that "human values" in the real world include war, slavery, religious fanaticism, racism, child abuse and one or two other inclinations of a similar ilk.

As noted by Barrot, renowned inventor Ray Kurzwell predicted that humans will eventually meld with AI technologies through cognitive enhancements to become "transhumanists." According to this theory, artificial general intelligence will ultimately evolve along with us.

According to Kurzwell, not only will computer implants enhance our brains' speed and overall capabilities, we will also be able to transport human intelligence and consciousness into computers. This strategy would ensure that super-intelligence will be at least partly human, controllable and "safe."

For instance, computer implants will enhance our brains' speed and overall capabilities. Eventually, we'll develop the technology to transport our intelligence and consciousness into computers. Then super-intelligence will be at least partly human, which in theory would ensure super-intelligence was "safe."

Barrot observes that a problem with this theory is that since even we humans aren't reliably safe, it is unwarranted to expect that super-intelligent humans will be either:

> *We have no idea what happens to a human's ethics after their intelligence is boosted. We have a biological basis for aggression that machines lack. Super-intelligence could very well be an aggression multiplier.*

Meanwhile, Barrot laments, 56 nations are developing battlefield robots, and the drive is to make them, and drones, autonomous. They will be machines that kill, unsupervised by humans.

Impoverished nations will be hurt most by autonomous drones and battlefield robots. Initially, only rich countries will be able to afford autonomous kill bots, so rich countries will wield these weapons against human soldiers of the impoverished nations.

Barrot asks us to imagine that in as little as a decade, a half-dozen companies and nations might field computers that rival or surpass human intelligence:

> *Soon we'll be sharing the planet with machines thousands or millions of times more intelligent than we are. And, all the while, each generation of this technology will be weaponized. Unregulated, it will be a catastrophe.*

Dire warnings about AI's dark side have been coming not only from the usual conspiracy theorists and technophobic Luddites, but also from the likes of tech icons Elon Musk, Bill Gates and Jeff Bezos.

"Mark my words," SpaceX and Tesla pioneer Elon Musk told attendees at the [2018] South by Southwest tech conference in Texas:

> *AI is far more dangerous than nukes. So why do we have no regulatory oversight? This is insane.*[cclxxxix]

Bioengineering our Humanity

If humans can invent machines which are increasingly smarter than we are, where does this lead? Are we in a sense "playing God" in a way that will render human reasoning obsolete? Can we integrate technological "thinking parts" into our biological anatomy to repair and replace failed sensory and motor response systems...just as we presently do with other organ and limb prosthetic devices?

Michael Bess, Vanderbilt University professor and author of *Our Grandchildren Redesigned: Life in a Bioengineered Society,* foresees a human future that is both terrifying and promising. In an interview with *Vox.com* contributor Sean Illing, Bess raises special concerns regarding AI's social influences on society and its ultimate potential to enable biological reengineering of our lives altogether.[ccxc]

Bess acknowledges that people have panicked about impacts of new information technologies since the invention of the printing press.

Even long before that, Socrates argued that reading a manuscript was nowhere near as insightful as talking with its author:

> *[Written words] seem to talk to you as though they were intelligent, but if you ask them anything about what they say, from a desire to be instructed, they go on telling you the same thing forever.*[ccxci]

Now, the advent of smartphones, computers and the Internet seem to be comparable in their impact to other big revolutions in communications and transportation that we've experienced over the past thousand years.

Bioengineering, however, is different. The impact of social media will pale in comparison to potential revolutions in AI or gene editing technologies. Bess projects that we're now on the verge of developing DNA-altering technologies that are so qualitatively different and more powerful that they will force us to reassess what it means to be human:

> *Bioelectric implants, genetic modification packages—the ability to tamper with our very biology—this stuff goes far beyond previous advances, and I'm not sure we've even begun to understand the implications.*

What's more, such capabilities are advancing at an unprecedented rate. Bass observes:

> *We went from having no World Wide Web to a full-blown World Wide Web in 20 or 25 years—that's astonishing when you consider how much the Internet has changed human life. In the case of, say telephones that took many decades to fully spread and become as ubiquitous as it is today.*

293

> *So what we've seen with the Internet is blisteringly fast compared to the past. For most of human history, the world didn't change all that much in a single lifetime. That's obviously not the case anymore, and technology is the reason why.*[ccxcii]

Bess worries that mankind doesn't have enough time to adapt to these changes. We'll need adequate time to alter our habits and to reappraise our cultural sense of who we are:

> *When these things happened slower in previous eras, we had more time to assess the impacts and adjust. That is simply not true anymore. We should be far more worried about this than we are.*

Sean Illing observed that our technology is developing much faster than our culture and institutions, and that this growing gap will eventually destabilize society. Still, here Bess was less pessimistic:

> *I think overall, as a society, we're insufficiently equipped, but that doesn't mean there aren't plenty of voices out there speaking sanity. What's interesting is that you can use these new technologies to get in touch with those voices and connect with other people who are questioning these technologies. The ability to connect in that way offers a lot of promise if it's used wisely.*

Professor Bess added that while many young people appear to

be walking around college campuses mindlessly staring at their phones, it's clear that even they understand what's happening and why it's problematic:

> *The more you live through screens, the more you're living in a narrow bandwidth, an abstract world that's increasingly artificial. And that virtual world is safe and controllable, but it's not rich and unpredictable in the way the real world is. I'm worried what will happen if we lose our connection to reality altogether.*

What's most striking about us humans, Bess observes, is that we are unpredictable in very basic ways. We're more complex than we can fathom, and there's something about us that is the opposite of artificial. It's the opposite of something made:[ccxciii]

> *All this genetic modification technology has the potential to take us into very worrisome territory where all the things we hold dear in our current world, all the values that give our lives meaning, are at risk. Either our survival is at risk or we become semi-machines who are like the marionettes of our own moment-to-moment experience. What becomes of autonomy? What becomes of free will? All these questions are on the table.*

Bess urges each of us to ask ourselves:

> *What does it mean for a human being to flourish? These technologies are forcing us to be more deliberate about asking that question.*

We need to sit down with ourselves and say, "As I look at my daily life, as I look at the past year, as I look at the past five years, what are the aspects of my life that have been the most rewarding and enriching? What things have made me flourish?

Bess concludes:

If we ask these questions in a thoughtful, explicit way, then we can say more definitely what those technologies are adding to the human experience and, more importantly, what they're subtracting from the human experience.[ccxciv]

Ratcheting up the bioengineering potentials even farther, what if artificial intelligence begets artificial life? Such an idea is no longer an implausible script of science fiction fantasy

As explained in the *Stanford Encyclopedia of Philosophy*, artificial life (Alife) is an outgrowth of AI technology to simulate or synthesize life functions:

The problem of defining life has been an interest in philosophy since its founding. If scientists were to succeed in discovering the necessary and sufficient conditions for life and then successfully synthesize it in a machine or through synthetic biology, then we would be treading on territory that has significant moral impact.[ccxcv]

One form of ALlife which was inspired by the work of

mathematician John von Neumann aims to achieve computational models which produce self-replicating cellular automata called "Loops." So far, these ALlife applications are content to create programs that simulate life functions rather than demonstrate "intelligence." A primary moral concern here is that these programs are designed to self-reproduce and in a way that resembles computer viruses. Successful Alife programs can potentially become computer malware vectors.

A second form of ALlife is based upon manipulating actual biological and biochemical processes in such a way that it produces novel life forms not seen in nature. It is much more morally charged.

In May 2010, scientists at the J. Craig Venter institute were able to synthesize an artificial bacterium called JCVI-syn1.0. Referred to as "Wet ALlife," this development tends to blur boundaries between bioethics and information ethics, potentially leading to dangerous bacteria or other disease agents, just as software viruses infect computers.[ccxcvi]

Some even argue that information is a legitimate environment of its own which possess intrinsic value that is in some ways similar to the natural environment, while in other ways foreign. Either way, they propose that information, as is on its own a thing, is worthy of ethical concern.

Therefore, if AI-directed robots are information machines, is it ethical to unplug and virtually kill them? Would HAL in *2001 Space Odyssey* really have cared?

Probably not.[ccxcvii]

Long before sci-fi author Arthur C. Clarke conceptualized the story of *2001 Space Odyssey*, Mary Shelley's famous 1818 novel, *Dr. Frankenstein* regretted meddling with nature. Will this same lesson prove to be the case with AI? Will our human story end in tragedy of Frankenstein proportions, or will we be able to live in harmony?

After all, much like many view AI today, even though the monster possessed moral and emotional sensibility, society unfairly and violently rejected its appearance and strength out of fear.

Despite good intentions and deeds, the poor creature just couldn't seem to win public support. As the monster described himself, "my life has been hitherto harmless and in some degree beneficial." He even used "extreme labour" to rescue a young girl from drowning, but no matter what he did, those actions were always misinterpreted. The public assumes that he was trying to murder the girl, and William Frankenstein even assumes that his monstrous creation plans to kill him.

Perhaps immodestly, the monster had a very good opinion of his superiority over his mortal detractors. He said:

> I was not even of the same nature as man. I was more agile than they and could subsist upon coarser diet; I bore the extremes of heat and cold with less injury to my frame; my stature far exceeded theirs.

Mary Shelley, the monster's real-life creator, understood our natural tendency to fear what we do not understand. She wrote: "Nothing is so painful to the human mind as a great and sudden change." As Victor Frankenstein lamented:

> I started from my sleep with horror; a cold dew covered my forehead, my teeth chattered, and every limb became convulsed: when, by the dim and yellow light of the moon, as it forced its way through the window shutters, I beheld the wretch—the miserable monster whom I had created.

Nevertheless, that fearsome creature conjured by Mary's imagination warned that unfairly pre-judged resistance to change would portend dire consequences. The monster cried out:

> *Shall each man find a wife for his bosom, and each beast have his mate, and I be alone? I had feelings of affection, and they were requited by detestation and scorn.*
>
> *Man! You may hate, but beware! Your hours will pass in dread and misery, and soon the bolt will fall which must ravish from you your happiness forever. Are you to be happy while I grovel in the intensity of my wretchedness?*
>
> *You can blast my other passions, but revenge remains—revenge, henceforth dearer than light or food! I may die, but first you, my tyrant and tormentor, shall curse the sun that gazes on your misery. Beware, for I am fearless and therefore powerful.*

Having been warned that acting against the monster's wishes would cause him to lose everything, including his good reputation, Victor recognizes that the danger to the world is greater than consequences to himself. Accordingly, he chooses to sacrifice himself to atone for his hasty rush into scientific inquiry.

Frankenstein-like theories regarding fears and fortunes of AI-driven monsters are subjects of contentious debate in today's scientific, technological, philosophical and public policy communities. Here, as in the past, our attention drifts to extremely contrasting and divided visions which are most dramatic rather than most likely.

One of the best-known members of the dystopian camp, Elon Musk, has called Super-intelligent AI systems "the biggest risk we face as a civilization," comparing their creation to "summoning the demon." Some sharing his view warn that when humans create self-improving AI programs whose intellect dwarfs our own, we will lose the ability to understand or control them.

Utopians, on the other hand, are more inclined to expect that once AI far surpasses human intelligence, it will provide us with near-magical tools for alleviating suffering and realizing human potential. Some holding this vision foresee that super-intelligent AI systems will enable us to comprehend presently unknowable vast mysteries of the Universe, and to solve humanity's most vexing questions such as eradication of diseases, natural resource depletion and world hunger.

Both of these scenarios would require that our AI developments lead to "artificial general intelligence" which can handle the incredible diversity of tasks accomplished by the human brain. Whether or not this will ever happen, much less how those tasks will be transformed and when, remain to be pure conjecture.

One of the Cassandras in the tech wilderness warnings about the Frankenstein-like threat of AI is author and speaker Sam Harris.

In an episode of *The Joe Rogan Experience,* he stated:

> You're talking about something that learns how to learn, in such a way that the learning transfers novel solutions—in the ultimate case, [it] can make improvements to itself.
>
> Once these machines become the best

> *designers of the next iteration of software and hardware, then you get this exponential take-off function, often called the singularity. There's a runaway effect where the capacities have gotten away from you.*
>
> *...[It's] not at all obvious to see a path forward that doesn't just destroy us.*

Pointedly, Max Tegmark, professor of physics at MIT, notes that humans used fire for quite a while before we figured out that we needed fire extinguishers.

Outsmarting Ourselves?

AI-enabled Internet of Things connectivity has forever changed our senses of personal space, our life and work opportunities, and countless aspects of our daily living routines. They have altered the ways we interact with loved ones, friends, employers and clients; have nurtured and supported a proactive business startup culture enabled through electronic commerce; and have opened up ways to identify and actively participate with groups and individuals throughout the world who share our special interests and problems.

Recent decades have witnessed the emergence of countless technological marvels with increasingly transformative influences on broad aspects of our daily lives. Examples include immediate access to the most recent information on virtually any subject; electronic conveniences that save personal and household time and money; enhanced mobility through shared on-demand transportation services that promise to banish most private automobiles to rusty scrap heaps of oblivion; and safety from predatory behaviors of others through ubiquitous, ever-watchful interconnected security devices. As elaborated in my book *The Weaponization of AI and the Internet,* expansively wired-together Internet of Things networks and a rapid emergence of "smart cities" openly invite autocratic control. We are therefore well-advised to be very cautious of all-too-seductive invitations to trade away precious privacy for promises of increased convenience, efficiency, and protection from those who are disposed to harm us.

A certain fact remains that AI, automation and the Internet are already impacting our lives in large and small ways that are legitimately argued as both good and bad. Moreover, the progeny of this triumvirate of Frankenstein monsters will continue to multiply to exert more and more influence over ever-broader aspects of our lives at an accelerating rate.

Lacking the apocalyptic drama of Hollywood blockbusters, some inherent perils might be likened by to a "boiling frog" analogy. Like a hapless frog placed comfortably in an open container of tepid water which is then brought slowly to a boil, it will not perceive danger until it is too late to jump out and is cooked to death.

Although gradual, there is broad recognition that smart technology will increasingly disrupt traditional structures of our social lives and economic livelihoods. Key among these impacts is a looming AI-driven work displacement crisis which will dramatically widen the wealth gap and pose a broad-spread challenge to maintenance of personal dignity.

Society has trained most of us to tie our personal worth to the pursuit of work and success. It will be painful for those who watch algorithms and robots replace them at tasks they have spent years mastering and proudly attending. Many will witness those tasks and entire industries disappear altogether, as ill-fated buggy whip manufacturers experienced following the invention of the internal combustion engine.

Joel Mokyr recognizes that while AI and automation are boosting economic growth by creating new types of jobs and improving efficiency in many businesses, they will be accompanied by negative effects on others. In the near-term, the less educated workers are likely to represent a disproportionate percent of job-loss casualties.[ccxcviii]

Dr. Kai-Fu Lee, author of *AI Superpowers: China, Silicon Valley and the New World Order,* argues that unprecedented disruptions applying existing AI technology to new problems will hit many white-collar professionals just as hard as it hits blue-collar factory workers. Still, he says:

> *Despite these immense challenges, I remain hopeful. If handled with care and foresight, this AI crisis could present an opportunity for us to redirect our energy as a society to more human pursuits: to taking care of each other and our communities. To have any chance of forging that future, we must first understand the economic gauntlet that we are about to pass through.*[ccxcix]

Lee points out that techno-optimists and historians would argue that productivity gains from new technology almost always produce benefits throughout the economy, creating more jobs and prosperity than before, but with a mixed bag of impacts. He notes:

> *The steam engine and electrification created more jobs than they destroyed, in part by breaking down the work of one craftsman into simpler tasks done by dozens of factory workers. But information technology (and the associated automation of factories) is often cited by economists as a prime culprit in the*

loss of U.S. factory jobs and widening income inequality.[ccc]

Lee concludes his article optimistically:

> *Artificial Intelligence will radically disrupt the world of work, but the right policy choices can make it a force for a more compassionate social contract.*

Finally, will tech overlords lord over all? Will those entities which control AI and information technologies ultimately determine society's winners and losers?

Critical uncertainties regarding threats posed by this new AI Frankenstein monster may revolve less around our mastery of its invention, or its mastery over us, and far more about who will ultimately master control over both.

Companies with more data and better algorithms will gain ever more users and data. This monopolistic self-reinforcing winner-take-all cycle will lead to God-like controls over all segments of society unknown in human history. Their instruments of power include dominion over information access and censorship, individual and business privacy, physical and economic security, transportation and energy infrastructures and financial levers of political influence which are growing at an astounding rate.

Recent events portend frightening global social consequences of amassing enormous quasi-government-level concentrations of wealth and power in a handful of monopolistic enterprises.

After concerns about Google's corporate practices lead to a revolt among the company's programmers, its CEO, Sundar Pichai, outlined new corporate guidelines for ethical principles.

Issuing a blog post, Pichai wrote that Google would not produce:

- Technologies that cause or are likely to cause overall harm.
- Weapons or other technologies whose principal purpose is to cause or directly facilitate injury to people.
- Technology that gathers or uses information for surveillance violating internationally accepted norms.
- Technologies whose purpose contravenes widely accepted principles of international law and human rights.

Pichai also laid out an additional seven principles to guide the design of future AI systems:

- AI should be socially beneficial.
- It should avoid creating or reinforcing bias.
- It should be built and tested for safety.
- It should be accountable. Incorporate privacy design principles.
- It should uphold high standards of scientific excellence.
- It should be made available for use.[ccci]

Google's ongoing behaviors directly contradict much of that lofty rhetoric.

In a speech on October 1, 2018 at the Hudson Institute, a conservative think tank focused on security and economic issues, Vice President Pence called on U.S. companies to reconsider business practices in China that involve turning over intellectual property or "abetting Beijing's oppression." He said,

"For example, Google should immediately end development of the Dragonfly app that will strengthen Communist Party censorship and compromise the privacy of Chinese customers." [cccii]

Mr. Pence's speech was the first public White House condemnation of Dragonfly, a mobile version of Google's search engine which is being designed and tested to adhere to China's strict citizen censorship program.

A spokeswoman for Google, a unit of Alphabet Inc. who declined to comment about the criticism, simply referred to a previous statement that described the company's work as exploratory and "not close to launching a search product in China." A logical follow-up question would be, "exploratory to what purpose?" [ccciii]

Chinese officials are known to be actively marketing their advanced facial recognition and other technologies to numerous other countries including Russia, Egypt, Turkey, and nations throughout Latin America. Ecuadorian law enforcement officials, for example, have purchased a network of Chinese security cameras with facial recognition software.[ccciv]

The globally wired-together Internet of Things now connects us through virtually every electronic device we own or use to vast networks of privacy-snooping and snitching data compilers and malevolent hackers. The range of such device vulnerabilities is vast and forever expanding. A small list of examples includes smart meters, thermostats, refrigerators, home security monitors, TVs, smart phones, web-connected cameras and wearable fitness trackers and health care devices.[cccv]

Aggressively marketed "smart city" proposals promise to make our lives more efficient and safer through ubiquitous Internet-connected and centrally monitored personal, household, municipal and regional systems which will be capable of constantly tracking our locations, our activities and our relationships with others.[cccvi]

Smart city messaging campaigns promote interconnected smart electrical meters, smart buildings, smart transportation services and smart citizen surveillance systems within smart grids. The idea is premised upon applying information and communication technologies (ICT) which collect and manage data on everything from air quality to noise levels to individual and group movements through an extensive network of sensors and surveillance cameras.

Smart city buyers beware! Such rhetoric energetically promulgated by big technology, engineering and consulting companies is predicated on the embedding of computerized sensors into the urban fabric so that bike racks and lamp posts, CCTV and traffic lights, remote-control air conditioning systems and home appliances all become interconnected into the wireless broadband Internet of Things.

The marketing campaigns are working. Global populations, Americans included, are trading away more and more of their personal privacy for promises of increased convenience and security.

Spy cameras are already sprouting up on lampposts and rooftops everywhere, facial recognition systems can track each of our individual movements and ICT and IoC networks are wiring private home appliances within municipal energy monitoring and eventual control networks overseen by George Orwell's Big Brother.

So here, once again, the discussion circles back to tradeoffs associated with delegating ever-more individual independence over our lives in exchange for promises of collective, wired-together efficiency and security benefits.

Harvard Law School Professor Lawrence Lessig providently predicted in 2000 that the Internet would become an apparatus that tracks our every move, erasing important aspects of privacy and free speech in our social and political lives. "Left to itself," he said, "cyberspace will become a perfect tool of control." [cccvii]

There are understandable reasons why citizens have increasingly come to be accept tradeoffs between gaining more security at the cost of less privacy in public venues. Many of us will recall footage from security cameras that cracked cases of the 2005 London subway and 2013 Boston Marathon bombings...and when Eric Cain was caught on camera shooting a Tulane University medical student named Peter Gold in 2015 (after Gold prevented him from abducting a woman on the streets of New Orleans).

Personal privacy becomes a much more urgent priority after it is surrendered. This loss is occurring at an incomprehensibly rapid and escalating pace.

Large data companies are already collecting data on their users, and until now, people had to be connected to their network to be seen. But now, those so-called "smart cities" are installing CCTV and other devices into panopticons where people can be watched every moment of their lives.

George Orwell's grim 1984 story admonition that "Big Brother is watching you" has since gained a rapidly growing number of progeny. That same year he wrote it—in 1949—an American company released the first commercially available CCTV system. Two years later, in 1951, Kodak introduced its Brownie portable movie camera.

Cities are rapidly expanding CCTV networks in the interest of public security. New York City ramped up installations following the September 2001 attacks to roughly 20,000 officially run cameras by 2018 in Manhattan alone. In 2018, Chicago reportedly had a network of 32,000 CCTVs to help combat the inner-city murder epidemic. Thanks to federal grants, Houston, which as recently as 2005 had none, had about 900 by 2018, with access to an additional 400.[cccviii]

An estimated 106 million new surveillance cameras are now currently being sold annually in the United States. Tens of thousands of cameras known as automatic number plate recognition devices, or ANPRs, hover over roadways to catch speeding motorists and parking violators.[cccix]

As marketed and presented, the data and the algorithms processing it are benign and indifferently neutral. We citizens who are constantly being observed and recorded are simply perceived as data points...information generators in various representative nodes in a system designed around the idea of data mining our ant-like patterns of behavior.[cccx]

This pervasive, ever-vigilant monitoring of our collective and individual activities and habits portends some very frightening implications regarding relinquishment of our privacy and prerogatives to invisible voyeurs and agenda-driven societal power-brokers who claim to represent our best interests.

As Songdo, South Korea researcher S.T. Shawayri points out, the data is never neutral, essential and objective in its nature. It is invariably "cooked" to recipe by chiefs embedded within institutions with aspirations and goals.[cccxi]

And it's beyond question that our Western nation attitudes have changed over the past century.

William Webster, a professor of public policy at the University of Stirling in Scotland, notes that the pre-World War II rhetoric about public safety was:

'If you've got nothing to hide, you've got nothing to fear.' In hindsight, you can trace that slogan back to Nazi Germany. But the phrase was commonly used, and it crushed any

sentiment against CCTVs.[cccxii]

Former U.K. deputy prime minister, Nick Clegg has observed:

> *And basically, it's happened without any meaningful public or political debate whatsoever. Partly because we don't have the history of fascism and nondemocratic regimes, which in other countries have instilled profound suspicion of the state. Here it feels benign. And we know from history, it's benign until it isn't.*[cccxiii]

Carnegie Mellon University professor of information technology Alessandro Acquisti has us remember:

> *In the cat-and-mouse game of privacy protection, the data subject is always the weaker side of the game.*

Acquisti reminds us that we in America haven't been through the experience of the man in the brown leather trench coat knocking on the door at four in the morning...so when we talk about government surveillance, the resonance is different. He warns:

> *[The desire for privacy] is a universal trait among humans, across cultures and across time. You find evidence of it in ancient Rome, ancient Greece, in the Bible, in the Quran. What's worrisome is that if all of us at an individual level suffer from the loss of privacy, society as a whole may realize its value only*

after we've lost it for good.[cccxiv]

As Gus Hosein, executive director of Privacy International, notes:

> *[If] the police wanted to know what was in your head in the 1800s, they would have to torture you. Now they can just find it out from your devices.* "[cccxv]

Electronic surveillance now follows us everywhere and from everywhere.

More than 1,700 satellites monitor our planet. From a distance of about 300 miles some can discern a herd of buffalo or stages of a forest fire.

Our skies have become cluttered with drones. About 2.5 million were reportedly purchased by American hobbyists and businesses in 2016. This doesn't include a huge fleet of unmanned aerial vehicles used by the U.S. government against terrorists and illegal immigrants.

Cameras connected with facial recognition are being used to by the Transportation Safety Administration (TSA) air marshals to trail and closely monitor unsuspecting Americans targeted for special airport inflight surveillance.

As reported by the *Boston Globe*, TSA's "Quiet Skies" program, which began in 2010, entitles and enables teams of undercover agents to document whether targeted individuals "change clothes or shave while traveling, abruptly change direction while moving through the airport, sweat, tremble or blink rapidly during the flight, use their phones, talk to other travelers or use the bathroom, among many other behaviors." [cccxvi]

The targets are not necessarily people who have done

anything that warrants any previous reasons to be on a terrorist watch list, although it is intended to identify those who "flag" reasons for concern. The first red flag is foreign travel—specifically, frequent visits to "countries that we know have a high incidence of adversarial actions."

Risk assessment targeting is far from reliable. The Globe reports that a flight attendant and a federal law enforcement officer are among those who have been flagged for surveillance under the program.[cccxvii]

Personal monitoring devices are proliferating...dash cams, cyclist helmet devices to record collisions, doorbells equipped with lenses to catch package thieves and inexpensive sound and movement-activated home security cameras are becoming ubiquitous.

As Rachel Holmes accurately observes in her Guardian.com article, it's now clear that the digital world has evolved into a creature of control. We realize that we are being watched, monitored and recorded, and we don't seem to care.

Facebook and other corporate Internet giants gather data to maximize profits from our consumer habits, from grocery shopping to TV viewing patterns to political interests and affiliations. Holmes writes:

> *Just like trawlers with dragnets, all sorts of other collateral data gets hauled in along the way. Data surveillance, once intangible and invisible, now blatantly announces its presence in our everyday lives. Mobile accessories and interconnectivity between gadgets and appliances in our homes—the Internet of Things—create an unprecedented network of tracking devices capturing data for commerce and government.*[cccxviii]

Holmes points out that the technology we thought we were using to make life more efficient started using us some time ago, and it's now attempting to reshape our social behaviors into patterns reminiscent of total surveillance culture. She writes:

> *In an increasingly online everyday life, our use of social media has become a medium for normalizing the acceptability of intrusion and behavioral connection. We are bombarded by 'helpful recommendations' on education, health, relationships, taxes and leisure matched to our tracked user profiles that nudge us towards products and services to make us better citizen consumers. The app told us that you only took 100 steps today. The ad for running shoes will arrive tomorrow.*

At the same time, we appear always ready to entrust that new technology with previously unimaginable judgment determinations that directly impact our lives and loved ones. Think self-driving cars, for example, where we will be asking computer algorithms to make split-second decisions to either save the car occupants or a school bus just ahead.

But then let's reconcile ourselves to recognizing that it will change anyway—and for many inevitably interconnected technological, cultural and economic reasons—just as societies and cities always have. Perhaps it's high time to get used to that idea.

In any case—and for better and worse—transformational changes will disrupt some populations more dramatically than others. I have predicted, for example, that deepening ideological divisions between predominately metropolitan and countryside populations will have profound national, state and local political consequences whereby regions of the United States with the largest cities will be disproportionately impacted by AI-driven demographic shifts.

Offers of guaranteed incomes to those whose jobs are displaced along with failed promises that government-distributed economic proceeds of AI and automation will liberate humans from labor. These guarantees will not sit well with country-wise Luddites who recognize these socialist agenda scams as exactly what they are.

To some experts, an AI world means more jobs, and more interesting ones; to others, it means a devastating loss of employment opportunities. To some, it means a deadly threat to human existence; to others, it means a better health and longer—perhaps much longer—lives.

To some, it means a time when AI can help us make smarter decisions; to others, it means the destruction of our privacy. While previous tech revolutions created jobs for unskilled workers, many or most of the new jobs that will be created by AI will require education and skills that most who lose their jobs lack.

The jobs that will be the most difficult to automate are those that require empathy and "people skills."

Perhaps most important, rather than replace jobs, robots and other AI systems will work alongside humans and enhance their knowledge and skills.

AI will also change many jobs beyond recognition. Truck driving, for example, is among those jobs at greatest risk once AI-powered autonomous vehicles hit the road.

The danger isn't from robots that will seek to control and destroy humans. No, it's more benign-sounding than that.

How might it happen? One possibility is that researchers succeed in creating a humanlike AI system—what is called artificial general intelligence, or AGI—that is capable of learning on its own and that could then design itself to be even more intelligent. In this event, which researchers refer to as the singularity, the machine could improve so rapidly that it turns into a superintelligence that is beyond our ability to monitor or control.

Quoting Stuart Russell, a professor of computer science at the University of California, Berkeley, it's almost impossible to anticipate every path a super AI might take to achieve its objective.

> *If you leave anything out, the AI system will find a way to take that thing you left out and shove it into infinity to help optimize the thing that you said you wanted.* "[cccxix]

Couldn't we just turn off a super-intelligent AI before it starts to do harm? Russell says that's not so easy:

> *...an AI that's hell bent on achieving its objectives would also realize that being shut down would prevent its ability to succeed and would try to stop any effort to pull the plug...like Hal in '2001: A Space Odyssey.'*
>
> *AI will become a constant companion. It won't be long before ii will be following us everywhere.*
>
> *The path to a ubiquitous AI isn't hard to imagine. AI is an all-pervasive, general-purpose technology, more like electricity than, say, an airplane. Like electricity, it eventually will be integrated into all aspects of our lives, homes, cars and offices, though in ways that are more*

*destructive and far-reaching. AI will drive us
to work in our autonomous cars, and once
we're there it will manage calendars, screen
and interview job candidates, run meetings,
and even take on some management tasks such
as forming work teams and assigning projects.*
*Back at home, smart devices will react
automatically to changing temperatures, noise
levels and air quality, change lights and music
to fit our moos and help children with their
homework.*

On the other hand, Luddite individualists and communities
won't eagerly trade in their freedom and convenience to travel
the countryside wherever and whenever they wish in private
vehicles in exchange for public transit tokens to destinations of
dwindling or remote interest. They will bristle at lapdog
attempts to turn virtually all roadways into tollways in order to
subsidize costly public transit networks that provide inefficient
and poor service in outlying suburban and rural areas. They will
continue to drive their children to school, load groceries in
their personal cars and carry building supplies and other cargo
in their vans and pickup trucks.

The Luddites will favor neighbor recognition over
electronic facial recognition in a community culture where
people look after one another out of view of privacy-intrusive
surveillance cameras. The Luddites will honor the importance
of civic participation over civil protest, and community values
that emphasize the importance of earned work ethics above
income-equality entitlement charities.

Luddite enterprise will prosper at the expense of lethargic
lapdog losses. Industries and businesses will follow work-
motivated labor workforces to more affordable and safer

countryside settings. Online entrepreneurship will flourish, including small work-from-home startups and retirement consulting. Online shopping and spacious land will attract more and more retail storage and outlet centers to serve growing populations. The demand for construction and maintenance services will grow, along with opportunities for a host of other skilled trades and professions.

Existing ideological and political tensions between lapdogs and Luddites will intensify. Lapdog anger will be inflamed over convictions that the Luddites are resisting the blessings and costs of technical nirvana promised by smart city efficiencies and comforts. The Luddites will resent and rebel against the lapdog culture of passivity, indolence and acquiescence to technology barons bearing costly socialist gifts they must pay for.

Which side will ultimately win and lose out in this contentious scenario? Very likely there will be some wins and losses both ways. Although I am admittedly rooting for the Luddites, perhaps there is still some final hope for the lapdogs as well. Maybe they will finally prove to be smarter than those hapless frogs sitting complacently in a pan while the water gradually heats to a boil after all.

Stephen Hawking has warned that artificial intelligent machines could wind up killing us all by themselves because they become too clever. Responding to a questioner during an "Ask Me Anything" session on Reddit, he replied:

> One can imagine such technology outsmarting financial markets, out-inventing human researchers, out-manipulating human leaders, and developing weapons we cannot even understand. Whereas the short-term impact of AI depends on who controls it, the long-term

impact depends on whether it can be controlled at all.[cccxx]

Finally, let's recognize that all heaven-or-hell scenarios such as any of these are like winning the Powerball jackpot...extremely unlikely.

Can we expect that future generations will witness angry protestors shouting "Hey hey, ho ho, AI overlords must go!"? Probably not.

Will smart tech ultimately outsmart humanity? No. Humanity can only surrender the dominion over its genius and the triumphant mastery over its inventions. Let's give our own marvelous intelligence, creativity and judgment more credit than that.

HUMANITY'S EXCEPTIONAL CHALLENGES

LET'S AT LEAST give ourselves credit for arriving and surviving so far.

After all, our ancestors made it through at least five major climate change, asteroid and volcanic events that variously killed between three-quarters and 96 percent of all life species: The Ordovician-Silurian Extinction (439 million years ago); the Late Devonian Extinction (between about 349-364 million years ago); the Permian-Triassic Extinction (about 251 million years ago); the Triassic-Jurassic Extinction (between about 119-214 million years ago); and the Cretaceous-Paleogene Extinction (about 65 million years ago).

Our early hominid family moved in to become apex predators after the last tyrannosaurus Rex dinosaurs checked out 65 million years ago, grew brain frontal lobes capable of self-awareness and tribal social behaviors and underwent a genetic mutation that allowed speech.

Within the past ten thousand years, our Sapiens ancestors competitively outlasted the Neanderthal hunter gatherers by inventing agriculture; battled and domesticated larger animals for food and clothing; established settlements, cities and empires, built great pyramids and cathedrals; formulated complex cultures and laws; developed advanced scientific methods and philosophies; and composed inspirational literature, music and sonnets.

Some inventive and adventuresome Sapiens contemplated the architecture and workings of a celestial universe and applied that knowledge to guide voyages of discovery, trade, conquest and migration to extend domains and dominions.

Others, within little more than the last century, have harnessed the power of lightning and atoms; mastered flight; travelled many times faster than the speed of sound; transmitted information from everywhere to everywhere else via orbital satellites; walked on the Moon and conceived artificial brains that can already outsmart their human creators.

And, with exceptional good fortune so far, we're now really only getting started.

Avoiding Self-Destruction

The development and use of terrifying technological tools for offensive military and disruptive civilian operations is already occurring at an escalating rate.

Imagine, for example, that everything begins as a perfectly ordinary day in your life. It is mid-afternoon and you are casually engrossed in an online Skype conversation with a friend regarding places to visit on a trip you plan to take.

Suddenly, you just as you hear a loud boom the screen on your laptop goes blank. You soon learn that the sound resulted from the blowout of a nearby power transformer.

Hopefully, someone will think to call it in for repair.

Strangely, however, cellphone and landline services are down as well.

In this scenario, and unbeknownst to you, frenzied turmoil is unfolding as federal, state, county and local government offices throughout the country are in a full blackout mode.

Darkness prevails everywhere, except where emergency generators provide isolated pockets of light and power. Soon, as batteries and fuel are exhausted, they will go out as well.

Hospitals are losing limited generator-supplied electricity to operate life-critical equipment. Meanwhile, primary power cannot be restored any time soon due to melted power transmission lines, damaged turbines at conventional power plants and lock-downs of nuclear plants.

Supplies of clean water are being depleted also. Metropolitan areas with high water towers atop high-rise buildings will have enough gravity flow to supply the most basic living needs for at best a few days. When that runs out, taps will go dry; toilets will no longer flush; emergency supplies of bottled water will become far too scarce for anything but drinking with no sources of replenishment.

Desperate food shortages soon occur due to distribution disruptions caused by rail system failures and jumbling of data at truck routing centers. Unable to withdraw cash from ATMs or bank branches, many Americans are being forced to go hungry, while others are looting stores. Police and emergency services engaged in rescuing people trapped in elevators have become overwhelmed. It seems that nearly everyone needs some kind of assistance.

Although most police are doing their level best to preserve calm and maintain civic order, they, like everyone else, lack access to critical information and adequate intercommunications means to coordinate responses.

There is no way of predicting when power and fuel outages which have now affect tens of millions of people over several states. While current fall weather conditions are fortunately moderate, many of those regions anticipate a coming winter home heating crisis which will put countless lives at risk.

There are few opportunities for people to leave for warmer climes. Public transportation, including trains, buses, and airlines are encountering the same fuel shortages that private motorists are. Gasoline and diesel fuel can't be delivered, and gas stations can no longer operate fuel pumps because there is no power to enable them to do so.

Amounts of water, food, and fuel consumed by those isolated in large metropolitan areas stagger comprehension. So do the growing mountains of uncollected waste, including human biological material, which has to create an unthinkable sanitary and health crisis.

There is also a rapidly- growing crisis of conviction that anyone in government is really in charge of solving this unmitigated disaster. A contagion of panic and chaos ensues that leads to one inevitable conclusion.

There never was a plan to deal with well-known realities that such a threat existed.

Everyone, every family, every community of friends and loved ones, is ultimately now on their own.

The previous scenario unabashedly borrows (or steals) extensively from premonitions put forth by two writers have advanced such truly ominous warnings.

The first source of these terrifying tidings is a 2015 book *Lights Out,* authored by former ABC news anchor and managing Nightline editor Ted Koppel.[cccxxi]

Koppel's dire premonition reportedly first arose in 2003 when a very large tree in Ohio had fallen on an electrical power transmission line. The event slowed down an overburdened control network resulting in cascading surges that tripped circuits providing electricity to 50 million people in eight states and two Canadian provinces. The blackout shut down everything that depended on the grid, including, for example, Cleveland's water system.

Fortunately, the outage lasted only a few hours for most people, and at most, a few days for others. And although a relief to many that it was later proven not to have resulted from a sinister cyberattack as many jittery people following the tragic 9/11 attacks two years earlier had suspected, it dramatically demonstrated how dependent we have all become upon power grid reliability.

A dozen years later, Koppel's book contemplated what could happen if a deliberate cyberattack on a US power grid were to knock out large power transformers and generators and render them irreparable. With few available spares, it could well require many months to create new ones and restore services.

Ted Koppel's 2015 book struck lots of critics as being overly alarmist. The notion that malevolent hackers could shut down a grid—much less cause permanent physical damage to critical equipment—seemed more like the script of a science fiction thriller like the 2015 television series *Madam Secretary* where an American President retaliates against Russia by plunging Moscow into darkness in winter.

The second source postulating potentials for such dire events is chronicled in the 2010 book *Cyber War,* co-authored by Richard Clarke and Robert K Knake. Clarke served in the White House as National Coordinator for Security, Infrastructure Protection, and Counterterrorism during the Ronald Reagan, George H.W. Bush, George W. Bush and Bill Clinton administrations.[cccxxii]

But could any of this scary hypothetical stuff happen in reality?

Ukraine, 2015:

It was the day before Christmas Eve, 2015, when many citizens in western Ukraine were very much in a religious, good-will holiday spirit.

That was just when the lights went out.[cccxxiii]

The timing had been ideal for saboteurs who hit Ukraine's power grid on that evening, and they weren't amateurs. They knew that the electric utility providers were operating with skeleton staffs, and for Vladimir Putin's army of patriotic hackers, Ukraine was a proven playground and testing haven for various sundry cyber spies, cyber vandals and cyber burglars.

It had been less than two years since Vladimir Putin had annexed Crimea and declared it would once again be part of Mother Russia. Putin's tanks and troops—who traded in their uniforms for civilian clothing and became known as the "little green men"—were busy sowing chaos in the Russian-speaking southeast of Ukraine. Their goal was to destabilize a new, pro-Western government in Kiev, the capital.

Cursors at Ukraine's Kyiv Oblenergo master control center suddenly began jumping across the screens as if guided by a hidden hand. Simultaneously, remote control hackers systematically disconnected circuits, deleted backup systems and shut down neighborhood substations.

Along with those disconnected circuits, the now-helpless operators' computer keyboards and mice had been disabled as well. The urgent situation was both bewildering and bizarre as if some paranormal powers had taken over their controls.

Putin's hackers also had other surprises for the controllers up their cyber sleeves.

The cheap malware program they had installed had been designed to permanently wipe out those control systems. Then, adding insult to injury, the hackers disconnected the Kyiv Oblenergo control room's backup electrical system, leaving the operators in little to do but curse in the dark.

Whereas Ukraine had suffered previous cyberattacks, this one was far worse. Power and computers across the country shut down, causing everything that depended upon them to fail. ATM's were closed, radiation safety monitors at the old Chernobyl nuclear plant went offline and automotive fuel pumps couldn't operate.

News broadcast stations were interrupted, and when they came back on the air found communications blocked with what appeared to be ransomware notices on their frozen computer screens with a Broken-English message informing users that their hard drives had been encrypted. The notice stated:

> *Oops, your important files have been encrypted...Perhaps you are busy looking to recover your files, but don't waste your time.*

Stealthily disguised to appear as a financial shakedown, users were told that their computer data could be unlocked by paying a $300 ransom in hard-to-trace Bitcoin cryptocurrency. The masquerade was entirely a ruse to offer the Russian government deniability cover.

Fortunately, the Ukraine cyberattack interrupted electricity for about 225,00 customers over a relatively brief period of a few hours. The impact would have been much worse had their grid not been controlled by old pre-computer-era metal switches in comparison with U.S. computer-dependent operations.[cccxxiv]

The "NotPetya" virus used to attack Ukraine was ultimately traced back to a hacking organization known as "Fancy Bear" run by the Main Directorate of the General Staff of the Russian Federation's military, often called the GRU.[cccxxv]

Estonia, 2007:

Russian hackers had been practicing and actively preparing for cyberattack hits on communist regime targets for at least a decade before Ukraine. Such a plan was perpetrated in the relatively small coastal Baltic city of Tallinn in Estonia on April 27, 2007.

Servers supporting Estonia's most-often utilized webpages suddenly became so flooded with cyber access requests that many systems collapsed under the load and shut down. Since known as a "distributed denial of service" attack—or "DDoS" for short—this strategy has become a major weapon in international cyber arsenals.

DDoS generates a pre-programmed flood of Internet traffic which employs a network of remote-control "zombie" computers called a "botnet" to overwhelm and crash the Internet- connected systems. Those zombie computers can either be programmed to patiently await remote orders or can immediately begin to seek other computers to attack.

Botnets generally begin by taking down websites. Then next target Internet addresses that most people would not know such as servers that run parts of telephone networks, credit card verification systems, and Internet directories.[cccxxvi]

Estonia, one of the most Internet-wired nations in the world, was particularly vulnerable to such attacks. Along with South Korea, the country ranks ahead of the United States in the extent of broadband penetration and broad public Internet utilization.

More than a million awakened computers became engaged in sending a flood of pings toward targeted Estonian servers.

Hansapank, the nation's largest bank, was targeted, spreading to other on-line commerce and communications networks nationwide.

Unlike most previous one-time DDoS assaults, this one repeatedly hit one site for a few days, then moved on to attack another. Ultimately hundreds which were hit week after week were unable to recover.

Cybersecurity experts who rushed in from Europe and North America used trace-back techniques to follow the pings to specific zombie computers, then watched to see when the infected machines "phoned home" to their masters. The messages were traced back to higher-level controlling devices in Russia.

Georgia, 2008:

The Republic of Georgia, a region geographically slightly smaller than South Carolina with a population of about four million people, lies directly south of Russia along the black sea. Viewed within the Kremlin's sphere of influence as a Russian territory, the two nations have had a contentious relationship dating back nearly a century.

In early August 2008, Ossetian rebels or Russian agents (depending upon whom you believe), staged a series of missile raids on some Georgian villages. The Georgian army retaliated by bombing the Russian-aligned South Ossetian capital city followed by invading the region on August 7.

Russian invasion foot soldiers were simultaneously joined by a small remote troop of cyber warriors whose primary mission was to prevent the isolated Georgians from knowing what was going on. They accomplished that goal by streaming DDoS attacks on Georgia media outlets and government websites along with access to outside CNN and BBC websites.[cccxxvii]

In preparing the cyber battlefield—before physical attacks—hackers had previously conducted more limited DDoS hits on Georgian government websites. Included was the webserver of the president's site.

The cyberattacks rapidly picked up intensity and sophistication. Routers that connect Georgia to the Internet through Russia and Turkey were flooded with so many incoming DDoS that outbound traffic was blocked.

Hackers also seized direct control of the rest of the routers supporting inbound traffic to Georgia. As a result, Georgia lost access both to outside news or information sources, and in addition, was denied means even to send an email out of the country.

As the Georgians attempted to accomplish "work-arounds" by shifting government websites to servers outside the country, the Russians countered every move. When they tried to block incoming DDoS traffic from Russia, the hackers rerouted their botnet attacks through servers in China, Turkey, Canada, and ironically, their former target victim, Estonia.

Georgia responded by transferring its president's government webpage to a server on Google's blogspot in California. As a further defense, the Georgian banking sector attempted to ride out the attacks by shutting down its servers altogether. The thinking was that it was better to suffer temporary financial online banking losses than to risk theft of critical data or damage to costly internal systems.

But the Russian hackers had prepared their work-around tricks for this occasion. Unable to penetrate Georgian banks, they deployed botnets that deliberately pretended to be a barrage of cyberattack traffic upon the international banking community originating *from* Georgia. Those attacks triggered automated Internet connections to the Georgian banking sector, paralyzing their operations. Credit card systems went down, soon followed by Georgia's mobile phone system.

The DDoS attack intensified and spread, ultimately enlisting barrages from six different botnets. The cyberassault infected and commanded computers from both unsuspecting and complicit volunteers. After downloading and installing software from numerous anti-Georgian websites, a volunteer could join the cyberwar by clicking on a button labeled "Start Flood."

Just as in Estonia, the Russian government incredulously claimed that the attack on Georgia was merely a populist activity that was beyond their control. Yet as pointed out by seasoned cyber expert Richard Clarke, any such large-scale activity in Russia, whether done by the government, organized crime, or citizens, is done with the approval of the intelligence apparatus and its bosses in the Kremlin.[cccxxviii]

America, Now:

In the fall of 2017, the U.S. Department of Homeland Security began to quietly warn American power grid companies of a likelihood that potential adversary nations were attempting to penetrate controls over their systems. This information was already well known to some of them that had already been monitoring and observing such efforts for years.[cccxxix]

Other grid companies were more complacent, dismissing the penetration risk as small since their controls were not connected to the internet. And besides, they argued that it was solely the US government's responsibility, not theirs, to protect them the citizenry from foreign attacks.[cccxxx]

By the summer of 2018, DHS's chief of industrial control systems analysis, Jonathan Homer, confirmed that Russia had successfully penetrated the US power grid. Homer reported:

"They [the Russian group, Dragonfly] have had access to the button, but they haven't pushed it."

Then-Director of National Intelligence and former Republican Senator Dan Coats described the Russian attacks and penetration of the U.S. electrical system being so severe that figuratively, "the warning lights are blinking red."

U.S. government officials explained that the Russian hackers had "jumped the air gap," the disconnection between internal grid control system networks and the Internet which provides an open information highway everywhere. In reality, however, few of those companies had isolated their controls from the Internet. Although many or most had segmented their internal networks by protective "firewalls," those precautions were seldom adequate to block penetrations by sophisticated hackers.

The Russians had dug into the grid infrastructure even deeper and wider, going after the companies that supply parts and do maintenance on the control side of the gap. Compromising those systems enabled attackers to tap into the log-in credentials of people with authorized access to the entire control network. This might allow hackers to remotely plug into the systems that display the state of the grid on monitors inside the control rooms to send false instructions and readouts to thousands of field devices.

Russia was not the only threat. By 2019, heads of all seventeen U.S. intelligence agencies were receiving confirmations that in addition to Russia's ability to disrupt the U.S. power grid, China could sabotage both the US power grid and natural gas pipeline system upon which the grid relies.

These are no longer theoretical threats. These are truly grim realities of today's new cyberworld.

Chinese government hackers are known to have

penetrated thousands of communications networks in the United States and tens of thousands around the world including numerous national embassies. U.S. intelligence officials have reported that China operatives have also inserted prospectively destructive malware into the U.S. power grid.

Iranian cyberattacks froze financial networks of the Bank of America and Chase and fried computers at the Sands Casino to demonstrate that a capacity for the ever-greater U.S. and global intimidation leverage.

North Korean hacks into the Sony Corporation, the Bangladesh Central Bank, and even U.S. and South Korean government websites are but sample harbingers of far the greater damage that even a small hermit rogue nation can wreak upon economically and technologically advanced powers.

Whereas the USA, China and Russia, were the first to invest significantly in building cyber warfare capabilities, several intelligence studies claim that more than 140 countries are now believed to be developing cost-effective but effective cyber weapons.[cccxxxi]

Early cyber events to date present potential previews of ominously impactful coming attractions. Such attacks can move at the speed of light, unlimited by geography and political boundaries. Being delinked from physics also means it can be in multiple places at the same time, meaning that the same attack can hit multiple targets at once.

Cyberattacks often differ from traditional attacks because instead of causing direct physical damage, they virtually always first invade another computer and steal the information within it. The intended results may be to damage something physical, but that damage always first results from an incident in the digital realm.

Military and civilian intelligence organizations, for example, routinely prepare cyber battlefields with virtual computer explosives in the form of software malware called "logic bombs" and "trapdoors" placing virtual explosives in other countries' power grids, financial and communications networks and other critical utility infrastructures.

Yes, and without any doubt, just as other governments do this to us, our cyber intelligence warriors do exactly the same to them. A joint U.S. and Israeli cyber sabotage of nuclear centrifuges in Iran also demonstrated technical capacities for other nations to shut down unambiguously vulnerable American power grids, energy pipelines and critical communications networks.

A big problem in our case, however, is that those nations with the greatest electronic infrastructure dependence also present richest, most vulnerable, cyberwar targets.

Every substantial cyberattack by one nation upon another provides a fair game precedent for potentially escalating see-saw levels of back-and-forth retaliation. A very considerable American disadvantage in this regard comes with living in the glassiest house in a very low-entry-cost stone fight.

Former White House security advisor Richard Clarke warns that at the very same time the United States prepares for offensive cyberwar, our current and continuing policies simultaneously make it increasingly difficult to defend against cyberattacks by aggressors. This is because we have simply too many vital networks to protect, and because additional ones are growing too quickly.

In simple terms, global offensive developments are wildly outpacing defense.[cccxxxii],[cccxxxiii]

Attempts to understand the new cyberwar era should begin by recognizing that rather than occurring "out there" in military domains, it fundamentally targets and involves everyone with a cellphone, laptop, car GPS device, or even—in some cases—a TV that they mistakenly didn't know was listening to their bedroom conversations.

In other words, it involves everyone and everything at every time that is connected to the Internet of cyberspace. That includes military intelligence, government communications, the facial recognition cameras on the lamppost, your home security system, and the smart meter that monitors your energy use.

In this broad but realistic frame of reference, cyberwarfare combat targets include all computers, the "Internet of Things," and social media communication platforms that engage the entire "sphere of human thought." These cyberattacks can and do cause both physical "kinetic" destruction of equipment as well as "non-kinetic" assaults on personal and proprietary data including intellectual property, financial systems and virtually everything in the realm of information, ideas and opinions.[cccxxxiv]

For all its countless benefits, this recent information and computational revolution have concomitantly spawned unfathomably terrifying cyberwar threats which might be comparable to World War II attempts to comprehend game-changing defense implications of the atomic bomb.

David Sanger, who like Richard Clarke teaches national security policy at Harvard's Kennedy School of Government, emphasizes fundamental differences between nuclear and cyberwar stratagems and consequences.

Professor Sanger points out that until the cyber age came along, America's two oceans symbolized our enduring national myth of invulnerability. The threat of nuclear attack preoccupied us during the Cold War, but generally, the United States has seemed confidently assured that it could take out dictators, conduct drone strikes on terrorists and blow up missile bases in faraway lands with little fear of retaliation.

After some terrifying close calls, notably the Cuban Missile Crisis in 1962, we found an uneasy balance of power to deter the worst threats with our primary adversaries through aptly labeled mutually assured destruction. Sanger notes that this common brutal understanding has worked thus far because the cost of failure on both sides is too high.

MAD stand-off nuclear threat deterrence between the United States and the Soviet Union seemed effective not only because each knew the other possessed world-destroying power, but also because each had confidence in the well-proven efficacy of its weapons systems.

Sanger believes that in the cyber age, we have not since MAD have opposing powers retained that strategic deterrence balance. Moreover, they probably never will because cyber weapons are entirely different from nuclear arms. He warns that although their effects have so far remained relatively modest, "to assume that will continue to be true is to assume we understand the destructive power of the technology we have unleashed and that we can manage it. History suggests that is a risky bet." [cccxxxv]

As Henry Kissinger wrote in his 1957 book *Nuclear Weapons and Foreign Policy*: "A revolution cannot be mastered until it is understood. The temptation is always to seek to integrate it into familiar doctrine; to deny a revolution is taking place."

It was time, he said, "to attempt an assessment of the technological revolution which we have witnessed in the past decade, to better understand how that revolution affected everything we once thought we understood." [cccxxxvi]

Terrifying Electromagnetic Impulses

In 2019, the U.S. *Congressional Commission to Assess the Threat from Electromagnetic Pulse Attack,* also known as the Congressional EMP Commission, released a previously secret report warning that Russia, North Korea, China and other nations are capable of launching devastatingly destructive attacks using relatively small, crude nuclear weapons exploded in space above America's mainland which—similar to a cyberattack—might cause many millions of casualties.

EMPs are pulses of energy that can be emitted from the blast of a nuclear weapon, portable devices like high power microwave weapons (HPMWs) or even certain natural phenomenon. These powerful pulses—when interacting with the Earth's magnetic field—have the ability to damage electronic and electrical equipment such as computers, cell phones, transformers and transmission lines, as well as critical communications infrastructure. Even worse, the design of America's electric grid means that damage to certain critical substations could cause cascading failures across the entire country.

The Congressional EMP Commission wrote in its July 2017 *Assessing the Threat from EMP Attack* report:

> *During the Cold War, the U.S. was primarily concerned about an EMP attack generated by a high-altitude nuclear weapon as a tactic by which the Soviet Union could suppress the U.S. national command authority and the ability to respond to a nuclear attack—and thus negate the deterrence value of assured nuclear retaliation.*
>
> *Within the last decade, newly-armed adversaries, including North Korea, have been developing the ability and threatening to carry out an EMP attack against the United States. Such an attack would give countries that have only a small number of nuclear weapons the ability to cause widespread, long-lasting*

damage to critical national infrastructures, to the United States itself as a viable country, and to the survival of a majority of its population.[cccxxxvii]

The report, written by Congressional EMP Commission chief of staff Peter Vincent Pry, projected that with ease of using a primitive nuclear weapon, a "New Axis" of aggressive nations could "black out" the Western world and dismantle all electricity and electronics, end water and food supply.

Such a blackout would seize the U.S. economy—causing disruptions among medical facilities, first responders, financial institutions, water and food distribution, communications networks and the transportation sector. A well-placed EMP would bring planes, trains and automobiles to a halt, and render our domestic military capabilities inoperable, as the Department of Defense relies on civilian infrastructure for 99% of its electricity needs. The consequences can potentially lead to the deaths of nine out of 10 Americans through starvation, disease and societal collapse.

Peter Pry reported to the *Washington Examiner* that EMP war plans combined with cyberwarfare are been known to have been drawn up by Iran, Russia, China, North Korea and even ISIS:

This new warfare uses cyber viruses, hacking, physical attacks, non-nuclear EMP weapons, and a nuclear EMP attack against electric grids and critical infrastructures. It renders modern armies, navies, and air forces obsolete. It paves the way for asymmetric warfare by small nations and terrorists.[cccxxxviii]

Pry said that the United States is an easy target because virtually everything, military and civilian, relies on computers, and even the Pentagon uses the civilian Internet:

> *Ours is the most technologically advanced society, and therefore the most susceptible to attack.*

In calling for greater Pentagon and Homeland Security attention to the issue, Pry compared the potential for an attack to Pearl Harbor. A nuclear explosion in the atmosphere above Omaha, Nebraska, for example, could black out Canada, the United States, and Mexico, causing "damage too broad and too deep to repair, requiring years, if the US could survive for years."

He said:

> *Although it is very difficult to predict exactly which electronic systems would be upset, damaged, or destroyed by an EMP attack, with certainty massive disruption and damage will be inflicted on unprotected electronics within the EMP field and, because of cascading failures, far beyond. EMP is analogous to carpet bombing or an artillery barrage that causes massive random damage that is specifically difficult to predict, but reliably catastrophic in its macro-effects.*[cccxxxix]

An *Executive Order on Coordinating National Resilience to Electromagnetic Threats,* signed by President Trump on March 26, 2019, recognizes that an EMP attack, in adversary military doctrine and planning, is a dimension of cyber warfare. The order states that the Assistant to the President for National Security Affairs (APNSA), working with the National Security Council and the director of the Office of Science and Technology Policy, "shall coordinate the development and implementation of executive branch actions to assess, prioritize, and manage the risks of EMPs." [cccxl]

Current efforts to protect the electric grid from an EMP event have ranged from hardening of infrastructure to updating technology and operational procedures. Duke Energy, for example, has begun testing its ability to recover from unexpected supply disruptions caused by an electromagnetic disturbance.

Human Race for Space

EMP dangers and defensive satellite-enabled threat monitoring and rocket intercept capabilities were spawned in a U.S.-Soviet Cold War race for military dominance. Similarly, the origins of all new technological "space era" marvels—both unthinkably terrifying and awesomely inspiring—were also spawned in ashes of wartime conflict. The invention of rocketry delivered death and carnage over World War II London skies.

Twelve years after President Truman authorized atomic bombings of Hiroshima and Nagasaki that expedited the end of that war, Soviet leader Nikita Khrushchev wasted no time pointing a nuclear finger back at the United States. On November 22, 1957 he told publisher William Randolph Hurst, Jr.:

> *The Soviet Union possesses intercontinental ballistic missiles. It has missiles of different systems for different purposes. All our missiles can be fitted with atomic and hydrogen warheads. Thus, we have proved our*

superiority in this area.

Even more pointedly, Khrushchev disparaged the defensive potency of U.S. naval power in a September 7, 1958 letter to President Eisenhower, stating:

> *In the age of nuclear and rocket weapons of unprecedented power and rapid action, these once formidable warships are fit, in fact, for nothing but courtesy visits and gun salutes, and can serve as targets for the right type of rockets.*[cccxli]

Revelations of Soviet ballistic missile advancements and geopolitical implications again made international headlines in 1962 when intended placements in Cuba of R-16 ICBMs triggered a fearsome Kennedy-Khruschev confrontation. Their range of about 2,200 kilometers combined with their basing in Cuba and western Russia posed a major threat not only to the United States, but also to bomber bases in Europe and Asia.

The international nuclear war threat subsided on October 27, 1962 when the show-down with Kennedy persuaded Khrushchev to order dismantling of the missiles in Cuba and their return to the USSR.[cccxlii]

The timing of two earlier events had also put great pressure on Kennedy to demonstrate resolute U.S. space leadership. A humiliating mid-April Cuban "Bay of Pigs" debacle and Yuri Gagarin's catapult into the space Hall of Fame during the previous month most certainly factored into the timing of Kennedy's May 25, 1961 announcement before a special joint session of Congress that an American astronaut would be safely sent to the Moon "before this decade is out."

The president also made it clear that this was to be a

competitive race dedicated to demonstrating U.S. technological supremacy over the Russians:

> *Within these last 19 months at least 45 satellites have circled the earth. Some 40 of them were made in the United States of America and they were far more sophisticated and supplied far more knowledge to the people of the world than those of the Soviet Union.*[cccxliii]

The high international prestige stakes of this competition weren't lost on Chief Designer Sergei Korolev and others in the Soviet Union.[cccxliv]

As quoted by Korolev associate Oleg Ivanovsky:

> *He would tell us that 'the Americans are at our heels, and the Americans are serious people.' He wouldn't use the word 'Amerikantsi' but 'Amerikan-ye' as if these weren't just American residents but the entire American culture we were competing with. He didn't mean this as an insult but as a show of respect for the competition.*[cccxlv]

The rest is truly human history.

Conceived in the ashes of war and deadly technical failures, we humans gave birth to a dream child that would mature to achieve inspirational goals, influencing the entire world in previously unimaginable and unquestionably beneficial ways.

Orbiting satellites have erased communication boundaries world-wide, have spawned a transformative internet information-sharing network, monitor natural and man-made events that affect our safety, coordinate and guide air and surface transportation movements and support unlimited international business opportunities.

Advancements in rocketry, spacecraft and instruments of exploration have opened an epic new era of cosmic pioneering discovery. And yes, the complex challenges driving such achievements have indeed yielded countless technological advancements and business opportunities that continue to enhance the quality of our everyday lives. In total, these advancements have expanded human experience while making our world seem smaller.

Such developments came as unforeseen results arising from a far different motivation, one prompted by events originating in the former USSR that harshly jolted the American psyche when a tiny Soviet satellite chirped alarming evidence of technological superiority. Soon afterwards an emboldened Soviet President Nikita Khrushchev banged that point home with his shoe on a table at a 1960 UN meeting, crowing "We will bury you."

Then, only one year later a young cosmonaut named Yuri Gagarin really rubbed it in, opening a new extraterrestrial era that threatened to leave the US behind.

America responded to the challenge.

Whereas Kennedy had set a national 1961 goal to put a human on the Moon and safely return him within a decade, America accomplished that and even more. By 1969 America had landed four of our citizens, plus delivered two more into lunar orbit who returned with them. Within three more years, eight others had walked on the Moon on successful round-trip voyages, along with four more Moon-orbiting companions.

As Kennedy warned at the time, these accomplishments didn't come easily. They required countless scientific and technological enablers, colossal coordination of complex operations, and exceptional dedication on the part of all those who dared to rise to the challenges.

The costs were enormous, including the lives of Gus Grissom, Ed White and Roger Chaffee who tragically perished during an Apollo command module ground test.

The Moon was only the beginning. The Skylab program (1973-79) established America's first true space station and demonstrated human abilities to adapt and undertake productive work under orbital weightless conditions.

Apollo-Soyuz enabled U.S. and Russian engineers and flight crews to rise above Cold War rivalry and work together in a literal high ground. American astronaut visitations to the Russian MIR space station extended this spirit of cooperation and diplomacy. Development and assembly of the International Space Station (ISS) now culminates the largest, most complex initiative in human history, a testament to great potentials and peaceful benefits of multi-national collaboration.

Yes, space exploration programs have produced technological innovations, but even more, they have served to stimulate interests in science and engineering-based studies, providing lessons and problem-solving challenges that apply at all levels of learning. The human conquest of space continues to inspire young and old to realize that the sky is literally no limit to what can be achieved with worthwhile goals, innovative vision and persistent commitment.

Pioneering New Worlds of Discovery

My close personal and professional long-time friend Buzz Aldrin emphasizes that we are at an important inflection point in human history. The decision is whether to look upwards and gain strength from vision and commitment to worthy goals beyond ourselves—beyond the here and now.

What Makes Humans Truly Exceptional?

Buzz urges that it is time that we sailed the sea of space once more with bold, expansive vision.

> *We must set our sights higher and be prepared to sail against the wind.*[cccxlvi]

Valiant strides forward in space will not only reflect our God-endowed greatness, but will summon us to make discoveries that, in turn, improve our lives on Earth.

Buzz and another friend, Neil Armstrong, were the first members of our species to leave human footprints on our neighboring lunar surface, followed by others I have had the great privilege to know. I will briefly note here that Neil served on the board of directors of Space Industries International, a company that I co-founded with the late former NASA Johnson Space Center Chief Engineer Maxime (Max) Faget and two other partners.

America and partnering nations now plan to again add many more footprints, to stay longer and to accomplish more.

The Moon, our closest celestial neighbor, offers a variety of interesting scientific features and potentially valuable resources. In addition to containing a record of planetary history, evolution and progress unavailable for study on Earth or elsewhere, recent detection of lunar ice has spurred growing international interest.

Orbiting Moon-circling instruments reveal that the Moon's North Pole alone may contain massive amounts of ice. This lunar-derived surface water can be used to support human activities. It can also be converted to liquid hydrogen and liquid oxygen fuel for future Mars-bound rockets.

The lunar surface is also known to contain enormously rare and valuable helium-3 to fuel future nuclear fusion reactors that can provide electricity not only on the Moon and Mars—

but on the Earth as well. Apollo 17 geologist/astronaut Harrison Schmitt, another personal friend and the last human to step on the Moon's surface, strongly advocates for these potentials.

Buzz believes that while the lunar surface can be used to develop advanced technologies, it is a poor location for homesteading...a lifeless, barren world of stark desolation. However, working with international partners, the global community can use the Moon as a testbed for tools, equipment and operations that will be needed for our next major destination...Mars.

Mars represents a new world of opportunity and discovery. Scientific and public interest in the planet has grown since 1960 telescope-driven observations have since been augmented by voyages of numbers of automated spacecraft sent there by multiple nations.

As Buzz chronicles:

> Mars has been flown by, orbited, smacked into, radar-examined, rocketed onto, bounced upon, rolled over, shoveled, drilled into, baked, and even laser-blasted. Still to come: Mars being stepped upon. The first footfalls will mark a historic milestone.

Buzz adds:

> While robotic exploration of Mars has yielded tantalizing clues about what was once a water-soaked planet and has revealed frozen water still trapped below the surface, the best way to study Mars is with the two hands, two eyes and two ears of a geologist. Humans surpass

*machines in speed, efficiency, nimbleness and
the dexterity to go places and do things. We
have the innate smarts, ingenuity and
adaptability to evaluate and respond in real-
time situations…to improvise, and to prevail
over surprises.*

*Recent and ongoing robotic exploration
of Mars is providing a window on a world that
can be home for new generations of colonists.
So yes, I suggest that going to Mars means
preparing for permanence on the planet.*

*Those who go to Mars need to have made the
decision to go there permanently. The more
people you have there, the more it can become
a sustaining environment. Except for very rare
exceptions, the people who go to Mars should
be prepared to remain.*

*Having them repeat their voyages, in my
view, is dim-witted. Why not allow them to
stay there? Did the pilgrims on the Mayflower
sit around Plymouth Rock waiting for a return
trip? They came to settle. And that's what we
should be doing on Mars.*[cccxlvii]

Permanent Mars habitation entails an entirely different goal
than the previous one of just putting people on the Moon. The
new pilgrims will require an ability to live off the land, a
circumstance that 102 other adventuresome souls once bravely
faced upon leaving England for a New World aboard a
Mayflower voyage. Martian settlers, however, will face much
stiffer challenges.

Much new research is needed to develop and test
technologies and methods to grow crops such as potatoes,

beans and wheat using methods that utilize on-site soil-derived water, oxygen and recycled wastes. This challenging area of study should involve scientific collaborations with Russia, China and other countries that are known to share this interest.

Buzz acknowledges that Mars settlement will inevitably invoke risks and casualties, just as other pioneering ventures have.

> *Unfortunately, pioneers will always pave the way with sacrifices. Over the decades, we have lost numbers of individuals—several of them close personal friends of mine—all intent on pushing the boundaries of exploration and seeking new horizons. Risk and reward is the weighing scale of exploring and taming space.*[cccxlviii]

As Buzz reflected in *No Dream is Too High,* untold numbers of people have experienced major failures and have recovered to become not only successful, but also as better, stronger people:

> *Failure is not a sign of weakness. It is evidence that you are alive and accepting of worthwhile risks.*
>
> *Successful innovators and doers who conceive things and make them happen combine awareness of failure risks of worthwhile enterprises with the willingness to take them. They are patient, resilient and don't quit. They experiment, often fail, learn more, and start over.*[cccxlix]

So finally, what does human spaceflight do for human society? Buzz reminds us that nothing is impossible if free people work together to accomplish great things:

> *Space exploration captures the imagination of our youth, it fuels the global workforce and economy with high technology jobs, and it fosters peaceful and beneficial international collaborations.*

Apollo 11 is a symbol of America at its best, people working together motivated by strong leaders with vision and resolve. In recognition of that legacy, Buzz now expresses great hope that a new generation of leaders and doers will once again boldly venture where no one has gone before.

> *American leadership in lunar exploration and development should be directed toward guiding other spacefaring nations in preparation for continuing activities that will lead to permanent human settlement of Mars. Earth and the Moon aren't the only worlds for us anymore.*
>
> *America needs to develop new strategies, technologies and spacecraft to bring us to the threshold of Mars by way of progressive missions to comets, asteroids, Mars' moons, and a permanent Mars settlement. In doing so, we fly by those comets and asteroids and sweep their surfaces to discover what the building blocks of the Universe are made of...step–by-step...just as Mercury and Gemini made Apollo possible. We move*

deeper into space in prelude to permanently
occupying the Red Planet.
    My friends, our Apollo days were a time
when we did bold things, achieving leadership.
Now is our time to be bold again in space.[ccc]

Exercising Exceptional Personal Potentials

As elaborated in my book *Thinking Whole,* one of the great blessings in my life has been to count many remarkably creative and innovative super achievers among my personal friends and working affiliations. Some of these individuals have gained prominent global and national distinctions, some have advanced truly important but less publicly recognized professional contributions and many, many more have quietly inspired my life and those of countless others as brilliant examples.

I have observed that the ability to deeply care about others is both our greatest human gift and reward.

Those who practice caring relationships develop a habit of bringing that character into all of their endeavors. Others trust them, and for good reasons. They have developed the capacity to focus their minds and energies upon helping to create a worthwhile result for everyone rather than concentrating on ways they can receive the greatest benefit, including claiming the most credit.

Truly caring people are willing to contribute more than their share without keeping a ledger to demonstrate that they deserve more. Often, they don't have to. Their generosity is usually recognized and valued by recipients who reward them far beyond their greatest expectations.

Through caring we experience life most fully. Caring provides motivations and rewards to love, seek, thrill, question, learn, solve, imagine, create, play, dare, feel, compete and commit.

Caring about self and others are not mutually exclusive. Both are vitally essential to our emotional and spiritual survival. Yet as Irish playwright, novelist and poet Oscar Wilde warned, dangers lurk in temptations to compromise our highest life expectations in order to achieve societal conformity:

> *I won't tell you that the world matters nothing, or the world's voice, or the voice of society. They matter a good deal. They matter far too much. But there are moments when one has to choose between living one's own life, fully, entirely, completely—or dragging out some false, shallow, degrading existence that the world in its hypocrisy demands. You have that moment now. Choose!*

Where can we find wellsprings of inspiration to raise us out of occasional doldrums, rekindle our enthusiasm about everyday events and encourage us to explore untested possibilities? We can begin by looking in familiar places: within ideas and examples of people we know that illuminate higher values to aspire to; among simple interests and pleasures that we take for granted and never fully recognize; and discover them connected to unresolved problems that cry out for innovative solutions.

As astrophysicist and science communicator Neil de Grasse Tyson observes:

> *The most successful people in life recognize they create their own love, they manufacture their own meaning, they generate their own motivation.*

The biggest issue isn't whether something can be accomplished, but rather, whether it should be, and who will do it. Mark Twain urged us to consider:

> Twenty years from now you will be more disappointed by the things you didn't do than by the ones you did do. So throw off the bowlines. Sail away from the safe harbor. Catch the trade winds in your sail. Explore. Dream. Discover.

Why do some people always seem to be at the right place at the right time and enjoy more opportunities than others do? Maybe it's because they realize that every place and time is right for something good to happen, and they are creative enough to figure out some possibilities. They are also objective enough to understand which options make the most sense, motivated enough to act upon them and perceptive enough to recognize when they don't and move on to something else.

Most opportunities don't have neon lights attached to them that flash the words "Here I am!" Instead, they are often camouflaged to blend into the background of everyday circumstances, or they are disguised to appear innocuous or even undesirable. Some opportunities have no form at all until new ideas shape them.

People who passively wait for opportunities to introduce themselves don't understand that the best ones usually don't behave that way. Being in popular demand, they seldom have to send out formal invitations, advertise in Want Ads or knock on doors to solicit interest. It takes a little bit of initiative to find them.

Really fine opportunities are personalized to fit our special interests and strengths. We're least likely to find these hanging around in settings that are dominated by assembly line mentalities—and which cater to one-size-fits-all aspirations. More often, they are discovered or created by individuals with discriminating opinions about who they are and what they expect. These people are selectively attuned to recognize possibilities that are most appropriate and accessible for them.

Opportunities frequently appear where they aren't anticipated and when they're not being sought. Sometimes finding them doesn't require any effort at all; we just open our minds, and there they are. This may seem unfair to people who work very hard, yet never seem to discover chances to get ahead. Perhaps they miss seeing them because their noses are too close to grindstones.

Big opportunities often come in deceptively small packages. Examples include chance meetings that introduce relationships and work prospects; satisfying hobbies that lead to new vocations; and casual observations that reveal exciting concepts to pursue.

These fortunate accidents can have profound influences over our lives. How can we prepare ourselves for these unforeseen developments? We can begin by realizing that everything we experience has potential importance.

The process of living is an aggregation of opportunities to experience, to grow, to contribute and to learn. So long as we're here, there's no good reason to miss out. Why not reach out instead?

The conditions and opportunities we recognize in life depend much more upon vision than location. Wherever we are, what we see is influenced by the viewing angles we choose and the ways we focus our minds. How we interpret what we see also reflects ways that we view ourselves in those settings.

We can choose to see our personal world any way we wish...both from the inside out and from the outside in. We also have the power to change it through our priorities and vision.

As Leonardo da Vinci reportedly observed:

> *It had long since come to my attention that people of accomplishment rarely sit back and let things happen to them. They went out and happened to things.*

Da Vinci also reportedly said:

> *I have been impressed with the urgency of doing. Knowing is not enough; Being willing is not enough; We must do.*

Sometimes survival is not nearly enough to wish for.

Consider the experiences of a small mollusk known as the sea squirt, for example. After swimming around early in life and eventually finding a permanent place such as a barnacle it can attach to, its thinking mission is then fully accomplished. The little squirt then absorbs its own brain which it no longer needs for nutrients to be rebuilt into other organs.

Even if no one, including ourselves, expects us to change the world, we all often do so in ways we can't fathom. For example, unless we live alone on islands, we affect lives of others just by being our natural selves.

We may intentionally or inadvertently introduce people to each other who fall in love and have families. We possibly influence people to relocate or change jobs as a result of our advice or assistance, causing their lives to take a whole new direction. And most assuredly, we touch and affect loved ones and friends in countless ways, large and small, on virtually a daily basis.

A lot of those interactive events—probably most—are

similar to "Brownian Motion," where, like molecules, we randomly bounce off others, transfer energy and change our own trajectories and theirs at least somewhat with nearly every contact. If these experiences are sometimes a bit bruising, they do help to keep things lively.

Although the global world may scarcely notice, we can also change our personal worlds by releasing limitations we perceive in ourselves. Buddha erased all boundaries for these potentials teaching that...

> *The mind is everything. What you think, you become.*

As Jefferson once asked (and answered):

> *Do you know who you are? Don't ask. Act! Actions will delineate and define you.*

William Shakespeare challenged us contemplate potentials far beyond our current selves:

> *We know what we are, but not what we may be.*

Or as George Bernard Shaw noted:

> *Life isn't about finding yourself, it's about creating yourself.*

Each of us has our own ideas about the sort of lives that we wish to experience. Imagining the sort of futures we want should be regarded as more than idle fantasies.

Dreaming up worthwhile futures can also energize us to

make them become real. Novelist Anais Nin, author of Delta of Venus, reminds us to remember:

> *We don't see things as 'they' are, we see them as 'we' are.*

If we don't visualize what we really want out of life, we won't have any basis for setting our course in the right direction in order to avoid wasting time and energy on routes destined to nowhere.

Dreaming about the future may be a forgotten art we must relearn from our inner child-selves in order to rediscover what we truly value most in ourselves and our lives.

And as Walt Disney proved:

> *If you can dream it, you can do it.*

As American Civil War abolitionist and humanist Harriet Tubman urges us to recognize:

> *Every great dream begins with a dreamer. Always remember, you have within you the strength, the patience, and the passion to change the world.*

History also reminds us that many of those people who have changed the world had to put their egos on the line and set disappointments behind them before their efforts were eventually rewarded.

President Abraham Lincoln urged us to "Always bear in mind that your own resolution to success is more important than any other thing." He drew upon personal experience on this matter, having previously been defeated before winning a seat in the Illinois legislature, then later being passed over twice in nomination bids for the U.S. Congress.

The late African anti-apartheid revolutionary leader Nelson Mandela encourages us to set our passion goals high. He advises:

> *There is no passion to be found in playing small—in settling for a life that is less than one you are capable of living.*

It is not selfish to expect everything possible in life. When we deny ourselves, we may also diminish our abilities to enrich others. In this regard, Mandela prompts us to make best advantage of opportunities afforded to constantly learn, grow, create and to transform passions into actions.

Have fun with your passions. Value them, nurture them and yes, apply them for good.

The art of "being" is a constantly evolving state of awareness and development; an open-ended pursuit of understanding; a perpetual process of "becoming." Opportunities for progress are retarded when we cling to fixed outlooks, intractable viewpoints and simplistic preconceptions that are falsely construed to be natural consequences of aging.

In reality, the opposite is true. Those limitations are causes, not results, of getting "old." We don't grow old. We become old when we allow ourselves to stop growing.

Growth and vitality, vs. stagnancy and obsolescence, are matters for personal definition. We risk missed opportunities leading to obsolescence whenever we pigeonhole ourselves into limiting self-concepts.

We can never be obsolete so long as we recognize our strengths and can find ways to apply them. And if we aren't entirely sure what those strengths are, then it's time to really start looking for challenging opportunities to discover them.

The longer these circumstances are permitted to continue, the more difficult it becomes to turn things around. Quoting a popular NASA space mission planning adage:

> *The sooner you fall behind, the more time you'll have to catch up.*

In putting off important changes, energy dissipates, confidence erodes, determination ebbs and hopes evaporate.

Or, as American writer, H. Jackson Brown Jr. observes in *Life's Little Instruction Book*:

> *Opportunity dances with those who are already on the dance floor.*

So, where do we find that inspiration to find worthwhile opportunities?

Sometimes it sneaks up on us unannounced when we're not looking, or we miss it when we search in the wrong places. Sometimes it shows up within ourselves and we reflect it on others.

However incomplete and inadequate our attempts to define it, inspiration is something needed to fill an otherwise human void.

It is something that guides our quest for understanding and practicing higher values. It is something that reveals forgotten beauty of nature and wisdom. It is something that provides examples of excellence we can aspire to and learn from, including generosity, courage, creativity, tenacity and true-life achievements. It is something that arouses our senses...something you feel when it touches you, sometimes prompting you to touch back.

Inspiration is all things we can imagine, and much, much more.

How do we recognize it?

Sometimes it arrives in our consciousness as a thunder clap of *WOW!...* or as a silent unexpected tear we shed when it softly touches our hearts. Sometimes it appears in the form of provident dreams upon which to construct marvelous thought castles of promise to house realities much larger than ourselves. Sometimes it is a force transmitted through bonds and connections of love and friendship that empower us...and often humble us as well.

Sometimes we are its agents. Without meaning to we inspire others through shared experiences and lessons. Sometimes inspiration transforms... other times it instructs. Sometimes it enriches a moment...at others, it influences a lifetime.

Oftentimes we let it inspiration find and recognize us.

Positive changes in our lives are often most fully enjoyed when we are free of preconceptions that limit clear vision; when we are ready for new challenges; when we are confident and optimistic about our abilities to make contributions; and when we are prepared to release restrictive tethers to the past.

Changing ourselves is often the most difficult challenge. It can require us to drop old and comfortable habits, embrace risks with uncertain rewards and abandon prejudices that filter the way we see things as well as the way others view us.

Changes offer us chances for fresh new starts and opportunities to reach higher experiences and goals. As John F. Kennedy noted:

> *Change is the law of life. And those who look only to the past or present are certain to miss the future.*

With so many choices available to us, it's often difficult to be certain which option is likely to turn out best.

As my friend David Eagleman and Anthony Brandt observe in their book, *The Runaway Species: How Human Creativity Remakes the World*:

> *Although separated by hundreds of millions of years, the brains of primitive creatures and corporate CEOs have the same questions to ask: how do I best balance exploiting my knowledge against exploring new territories?*
>
> *No creature, or business, gets to rest on the laurels of past success: the world changes unpredictably. The survivors are those who stay dexterous, responding to new needs and opportunities.*[cccli]

We inevitably face situations which require us to make very difficult decisions that catch us completely off guard and unprepared with little time or information to assess options and implications. Some of the choices may involve high-stakes gambles that we can't avoid, where lives, relationships or financial survival might hang in the balance.

The really tough decisions that we make reveal our true character. They force us to confront our worst fears, come to terms with fundamental values and take charge in the face of great uncertainties. Throughout history, our human responses to these challenges have motivated our ancestors to learn from failed and successful experiences, to adapt and ultimately, to survive.

Quoting English poet, W.H. Auden Markings:

> *I am sure it is everyone's experience, as it has*

*been mine, that any discovery we make about
ourselves or the meaning of life is never, like a
scientific discovery, a coming upon something
entirely new and unsuspected; it is rather, the
coming to consciousness recognition of
something, which we really knew all the time
but, because we were unwilling to formulate it
correctly, we did not hitherto know we knew.*

I will now conclude this book by admitting that the time and thought devoted to this project has been undertaken as a wonderfully selfish and personal mind-expanding adventure.

This thinking and writing journey began with a fundamental premise...namely, that through releasing self-imposed limitations about who we are both as individuals and members of an exceptional human community frees us to embrace, explore and experience more fully enriched lives.

The essential "selves" that form our sense of human and personal identity, our aspirations, our inspirations and our progress towards realizing the sort of lives we imagine we most desire are a perpetual work in progress. We become more or less the truly exceptional people we wish to be become through practice and habit.

Whether that person we are constantly in the process of becoming is the same exceptional individual we most admire and wish to emulate depends a lot upon whether we believe it is worth all the effort necessary to take on that burdens of opportunity and responsibility.

The persistent will to succeed—to grow—to contribute—to innovate—reflects a large capacity to care deeply enough about somethings and someones to remain passionately energized and unflappably determined. The greatest rewards of caring come when our focus goes beyond immediate self-gratification and comfort.

I will sadly conclude this gratifying contemplation adventure with five major take-away lessons drawn from my own experiences that I previously summarized in *Thinking Whole:*

Lesson One is that if you can't sit still, then really get moving! You can't rely entirely upon the swift current of events to carry you safely through all of the obstacles. Instead, you sometimes have to paddle like hell to take command of your options so that they aren't left to chance or fall under the control of others.

People who constantly operate in a reactive mode are typically overwhelmed, confused and ineffective. Even though you may not have all of the desired knowledge and confidence, it is important to set some form of purposeful, yet flexible plan into action. Put yourself in charge of your own choices. They are too vital to delegate.

Lesson Two is to make it a practice to look at the *BIG PICTURE* so that you don't get lost in the details. In your personal life, try to be clear about what it is that you care most about. In a new job or position assignment, work to understand the broad priorities and scope of the business or service environment of the organization and the ways that your designated role fits into and contributes to them.

Constantly strive to visualize your participation within the context of larger purposes and processes which will be impacted by your performance. Focus on things that are most important and make certain that they are accomplished well and on time. Get and stay organized.

Lesson Three is to define your opportunities, responsibilities and prerogatives as ambitiously as possible. Push the boundaries of your personal expectations and job descriptions to the limit. Be more than anyone expects you to be.

Don't worry about what others might consider to be either beneath or above your appointed station. Find ways to help friends and associates look good and be an advocate for them. Think of any title or status that you have as an opportunity to take action rather than as a license to take credit. See yourself as a leader, and act like one.

Lesson Four is to value your own resources and abilities. If you don't know how to solve a problem someone else's way, then devise your own approach. Don't allow yourself to be intimidated by things that you don't immediately comprehend merely based on a lack of background information and experience. Just absorb what you can and make it a point to later investigate points that you missed.

Learn to be patient with yourself. Don't be afraid to ask questions, but do this thoughtfully, especially when formal occasions warrant some discretion.

Heed Henry Ford's advice: "Whether you believe you can do a thing or not, you are right."

Lesson Five is to learn from others but be yourself. Everyone has their own unique strengths and styles. When you attempt to emulate others or compete with them on their terms, you are likely to fail. Doing that often tends to cloud your awareness of your own unique qualities, impairs natural creativity and keeps you off balance. Besides, being who you are can be a lot of fun once you get the hang of it.

You can't always buck the current and reverse direction, or even steer to a safe refuge on shore. So, you better learn to swim, get ready for that raging white water, do your best to stay clear of the boulders and submerged logs and be prepared to get your feet wet.

Dive in and enjoy the thrill.

And yes, become even more truly exceptional.

---

[i] *What made the Big Bang bang?,* Alan Guth and Neil Swidley, May 2, 2014, Boston Globe Magazine
http://www.bostonglobe.com/magazine/2014/05/02/alan-guth-what-made-big-bang-bang/RmI4s9yCI56jKF6ddMiF4L/story.html
[ii] Encyclopedia of Death and Dying,
http://www.deathreference.com/Py-Se/Reincarnation.html
[iii] *Refugees for life in a hostile Universe,* Guillermo Gonzalez, Donald Brownie, and Peter D. Ward, Scientific American, 2001
[iv] *Extinction: Bad Gene or Bad Luck?,* David M. Raup, New York: W.W. Norton & Company, 1992
[v] *Wonderful Life: The Burgess Shale and the Natural History,* Stephen J. Gould, New York: W.W. Norton 7 Company, 1989
[vi] *Mystery of Mysteries: Is Evolution a social Construction?,"* Michael Ruse, Cambridge, Massachusetts: Harvard University Press, 1999.
[vii] *On the difficulties of making Earth-like planets,* Stuart Ross Taylor, The Leonard Award Address, Meteorites & Planetary Science, V.34, pp.317-29, 1999.
[viii] *Mapping Human History: Discovering the Past through Our Genes,* Steve Olson, Boston, Houghton Mifflin Company, 2002.
[ix] *Shadows of Forgotten Ancestors: A Search for Who We Are,* Carl Sagan and Ann Druyan, New York: Ballantine Books, 1992.
[x] *Out of Chaos, Evolution from the Big Bang to Human Intellect,* Wayne M. Bundy, 2007, Universal Publishers.
[xi] *Life and Death of Planet Earth,* Peter Ward and Donald Brownlee, Rare Earth, 2003.
[xii] *Out of Chaos, Evolution from the Big Bang to Human Intellect,* Wayne M. Bundy, 2007, Universal Publishers.
[xiii] *Life and Death of Planet Earth,* Peter Ward and Donald Brownlee, Rare Earth, 2003.

[xiv] *The Ages of Gaia: A Biography of Our Living Earth,* James Lovelock, edited by Lewis Thomas, New York: W.W. Norton & Company, 1995.
[xv] *The River of Consciousness,* Oliver Sacks, New York-Toronto: Alfred A. Knopf, 2017.
[xvi] Ibid.
[xvii] Ibid.
[xviii] Ibid.
[xix] Ibid.
[xx] Ibid.
[xxi] *The Brain: The Story of You,* David Eagleman, New York, Vintage Books, A division of Random House, LLC, 2015.
[xxii] Ibid.
[xxiii] Thinking Whole: Rejecting Half-witted left & Right Brain Limitations, Larry Bell, Stairway Press, 2018.
[xxiv] Ibid.
[xxv] *A Theory of Therapy, Personality Relationship as Developed in the Client-Centered Framework,* Carl Rogers. In (Ed.) S Koch, Psychology: A Study of Science, Vol. 3: Formulations of the person and the social context, New York: McGraw Hill, 1959.
[xxvi] *The Human Advantage,* Suzana Herculano-Houzel, Cambridge, London: The MIT Press, 2016.
[xxvii] *The Dragons of Eden: Speculations on the Evolution of Human Intelligence,* Carl Sagan, New York: Random House Publishing Group, Ballantine Books, 1977.
[xxviii] The Ancestor's Tale: A Pilgrimage to the Dawn of Evolution, Richard Dawkins, Boston: Houghton Mifflin Company, 2004.
[xxix] *"The Double Helix,"* James D. Watson, New York: Atheneum, 1968.
[xxx] *The Ages of Gaia: A Biography of Our Living Earth,* James Lovelock, edited by Lewis Thomas, New York: W.W. Norton & Company, 1995.
[xxxi] *Out of Chaos, Evolution from the Big Bang to Human Intellect,* Wayne M. Bundy, 2007, Universal Publishers.

[xxxii] *An Ancestor to Call Our Own,* Kate Wong, Scientific American, June 2003.

[xxxiii] *The Dragons of Eden: Speculations on the Evolution of Human Intelligence,* Carl Sagan, New York: Random House Publishing Group, Ballantine Books, 1977.

[xxxiv] *Shadows of Forgotten Ancestors: A Search for Who We Are,* Carl Sagan and Ann Druyan, New York: Ballantine Books, 1992.

[xxxv] *Out of Chaos, Evolution from the Big Bang to Human Intellect,* Wayne M. Bundy, 2007, Universal Publishers.

[xxxvi] *Civilization left its mark on our genes,* Bob Holmes, New Scientist, December 24-January 6, 2006.

[xxxvii] *The Role of the Prefrontal Cortex in Dynamic Filtering,* Ralph Adolphs, Psychology, 2000.

[xxxviii] *The Evolution of Imagination,* Stephen T. Asma, University of Chicago Press, 2017.

[xxxix] Ibid.

[xl] Neocortex Size as a Constant on Group Size in Primates, Robin Dunbar, Journal of Human Revolution, 1992.

[xli] *Understanding Primate Brain Evolution,* R.I.M Dunbar and Susanne Shultz, Philosophical Transactions of the Royal Society of London B: Biological Sciences, 2007.

[xlii] *Our brains they are a-changing,* Mason Inman, New Scientist, September 17-23. 2005.

[xliii] *Emotional Intelligence: Why it Can Matter More than IQ,* Daniel Goleman, New York: Bantam Books, 2006.

[xliv] *Shadows of Forgotten Ancestors: A Search for Who We Are,* Carl Sagan and Ann Druyan, New York: Ballantine Books, 1992.

[xlv] *When Did Man Discover Fire?: Ancestors of Modern Humans Used Fire 350,000 Years Ago,* New Study Suggests, International Business Times, December 15, 2014.

[xlvi] *Sapiens: A Brief History of Humankind,* Yuval Noah Harari, New York, London, Toronto, Sydney, New Delhi, Auckland, Harper Perennial, 2015.

[xlvii] *Out of Chaos: Evolution from the Big Bang to Human Intellect,* Wayne Bundy, Boca Raton: Universal Publishers, 2007.

[xlviii] *The Road to Now: Taking Stock of Evolution and Our Place in the World*, Melvin Bolton, Allen & Unwin: Crows Nest NSW, Australia, 2001.

[xlix] *Sapiens: A Brief History of Humankind,* Yuval Noah Harari, New York, London, Toronto, Sydney, New Delhi, Auckland, Harper Perennial, 2015.

[l] *Out of Chaos: Evolution from the Big Bang to Human Intellect*, Wayne Bundy, Boca Raton: Universal Publishers, 2007.

[li] *The Substance of Civilization: Materials and Human History from the Stone Age to the Age of Silicon*, Stephen Sass, New York: Arcade Publishing, 1998.

[lii] *Before the Dawn: Recovering the Lost History of Our Ancestors*, Nicholas Wade, New York: The Penguin Press, 2006.

[liii] *Sapiens: A Brief History of Humankind,* Yuval Noah Harari, New York, London, Toronto, Sydney, New Delhi, Auckland, Harper Perennial, 2015.

[liv] *The Prehistory of the Mind,* Steven Mithen, London: Thames and Hudson, 1999.

[lv] *Out of Chaos: Evolution from the Big Bang to Human Intellect*, Wayne Bundy, Boca Raton: Universal Publishers, 2007.

[lvi] *Wisdom of the West,* Bertrand Russell, London: Bloomsbury Books, 1989.

[lvii] *The Age of the Spiritual Machines*, Ray Kurzweil, Viking, 1999.

[lviii] *Sapiens: A Brief History of Humankind,* Yuval Noah Harari, New York, London, Toronto, Sydney, New Delhi, Auckland, Harper Perennial, 2015.

[lix] Ibid.

[lx] *Wisdom of the West,* Bertrand Russell, London: Bloomsbury Books, 1989.

[lxi] *The Evolution of Imagination: An Archaeological Perspective*, Steven J. Mithen, SubStance 30, 2001.

[lxii] *Shells of the French Aurignacian and Perigordian,* Yvette Taborin, in *Before Lascaux: The Complete Record of the Early Upper Paleolithic*, ed. Heidi Knecht, Anne Pike-Tay and Randall White, Boca Raton: CRC Press, 1993.

[lxiii] *Sapiens: A Brief History of Humankind,* Yuval Noah Harari, New York, London, Toronto, Sydney, New Delhi, Auckland, Harper Perennial, 2015.

[lxiv] *Double Child Burial from Sunghir (Russia): Pathology and Inferences for Upper Paleolithic Funerary Practices,* Vincenzo Formicola and Alexandra P. Buzhilova, American Journal of Physical Anthropology, 2004.

[lxv] *Sapiens: A Brief History of Humankind,* Yuval Noah Harari, New York-London-Toronto-Sydney-New Delhi-Auckland: Harper Perennial, 2015.

[lxvi] *The Prehistory of the Mind: A Search for the Origins of Art, Religion and Science,* Steven Mithen, London: Thames and Hudson Ltd., 1996.

[lxvii] *The Evolution of Imagination,* Stephen T. Asma, University of Chicago Press, 2017.

[lxviii] *The Age of Insight: The Quest to Understand the Unconscious in Art, Mind and Brain: From Vienna to the Present,* Eric Kandel, Random House, 2012.

[lxix] *The Evolution of Imagination,* Stephen T. Asma, University of Chicago Press, 2017.

[lxx] *An Early Bone Tool Industry from the Middle Stone Age at Blombos Cave, South Africa; Implications for the Origins of Modern Human Behavior, Symbolism and Language,* C.S. Henshilwood, F. d'Errico, C.W. Marean, R.G. Milo, and R. Yates, Journal of Human Evolution, 2001.

[lxxi] *Sapiens: A Brief History of Humankind,* Yuval Noah Harari, New York, London, Toronto, Sydney, New Delhi, Auckland, Harper Perennial, 2015.

[lxxii] *The Evolution of Imagination: An Archaeological Perspective,* Steven J. Mithen, SubStance 30, 2001.

[lxxiii] *The Origins of Greek Thought,* Jean-Pierre Vernant, New York: The Penguin Press, 1984.

[lxxiv] *Behavioral and Neural Correlates of Delay of Gratification 40 Years Later,* B.J. Casey et al., Proceedings of the National Academy of Sciences, 2011.

[lxxv] *The Substance of Civilization: Materials and Human History from the Stone Age to the Age of Silicon*, Stephen Sass, New York: Arcade Publishing, 1998.

[lxxvi] *The Artful Universe: The Cosmic Source of Human Creativity*, John D. Barrow, Boston: Back Bay Books, Little Brown, and Company, 1998.

[lxxvii] *The World Economy, Vol. 2*, Angus Maddison, Paris: Development Centre of Organization of Economic Cooperation and Development, 2006.

[lxxviii] Sapiens: A Brief History of Humankind, Yuval Noah Harari, 2015, HarperCollins, Harper Perennial.

[lxxix] *A History of Knowledge: Past, Present, and Future*, Charles Van Doren, New York: Carol Publishing Company, 1991.

[lxxx] *The Lost Civilizations of the Stone Age*, Richard Rudgley, New York: Simon & Schuster, 2000.

[lxxxi] *Sapiens: A Brief History of Humankind*, Yuval Noah Harari, 2015, HarperCollins, Harper Perennial.

[lxxxii] Ibid.

[lxxxiii] *Martyrdom and Persecution in the Early Church*, W H Frend, Cambridge: James Clarke & Co., 2008.

[lxxxiv] Sapiens: A Brief History of Humankind, Yuval Noah Harari, 2015, HarperCollins, Harper Perennial.

[lxxxv] *Why I am a Hindu*, Shashi Tharoor, New Delhi: Aleph Book Company, 2018; as summarized in the Wall Street Journal, *How Hinduism Has Persisted for 4,000 Years*, January 18, 2019.

[lxxxvi] *Axialism and Empire*, Sheldon Pollock, in Axial Civilizations and World History, ed. Johann P. Arnason, S.N. Eisenstadt and Bjorn Wittock, Leiden: Brill, 2005.

[lxxxvii] *China: A History*, Harold M. Tanner, Indianapolis: Hackett Publishing Company, 2009.

[lxxxviii] *Guns, Germs and Steel: The Fates of Human Societies*, Jared Diamond, New York: W.W. Norton and Company, 1999.

[lxxxix] *The Genius of China: 3,000 Years of Science, Discovery, and Invention*, Robert Temple, New York: Simon and Schuster, 1986.

[xc] *Doubt and Certainty,* Tony Rothman and George Sudarshan, Reading: Perseus Books, 1998.

[xci] *Connections,* James Burke, Boston: Little, Brown and Company, 1995.

[xcii] *The Passion of the Western Mind: Understanding the Ideas that Have Shaped Our World View,* Richard Tarnas, New York: Ballantine Books, 1991.

[xciii] *The Passion of the Western Mind: Understanding the Ideas that Have Shaped Our World View,* Richard Tarnas, New York: Ballantine Books, 1991.

[xciv] *Wisdom of the West,* Bertrand Russell, London: Bloomsbury Books, 1989.

[xcv] Ibid.

[xcvi] *The Origins of Greek Thought,* Jean-Pierre Vernant, Ithaca: Cornell University Press, 1984.

[xcvii] *Wisdom of the West,* Bertrand Russell, London: Bloomsbury Books, 1989.

[xcviii] *The Origins of Greek Thought,* Jean-Pierre Vernant, Ithaca: Cornell University Press, 1984.

[xcix] *Wisdom of the West,* Bertrand Russell, London: Bloomsbury Books, 1989.

[c] Ibid.

[ci] *Cleopatra & Antony,* Brian Haughton, Ancient History Encyclopedia (article), January 10, 2011.

[cii] Ibid.

[ciii] *The Passion of the Western Mind: Understanding the Ideas that Have Shaped Our World View,* Richard Tarnas, New York: Ballantine Books, 1991.

[civ] *The Crusades: Motivations, Administration, and Cultural Influence,* Rachel Rooney with Andrew Miller, Digital Collections for the Classroom, The Newberry.org, October 10, 2016.

[cv] *The History of the Rise and Fall of the Roman Empire,* Edward Gibbon, (Six volumes), London: Strahan & Cadell, 1776-1789.

[cvi] *History of the Byzantine Empire,* Alexander Vasiliev, University of Wisconsin Press, 1952.

[cvii] *Wisdom of the West,* Bertrand Russell, London: Bloomsbury Books, 1989.

[cviii] *Landmarks in Western Science: From Prehistory to the Atomic Age*, Peter Whitfield, New York: Rutledge, 1999.

[cix] *The Passion of the Western Mind: Understanding the Ideas that Have Shaped Our World View,* Richard Tarnas, New York: Ballantine Books, 1991.

[cx] *What the History Books Left Out,* Ehsan Masood, New Scientist, April 1-7, 2006.

[cxi] *The Passion of the Western Mind: Understanding the Ideas that Have Shaped Our World View,* Richard Tarnas, New York: Ballantine Books, 1991.

[cxii] *Landmarks in Western Science: From Prehistory to the Atomic Age*, Peter Whitfield, New York: Rutledge, 1999.

[cxiii] *The Passion of the Western Mind: Understanding the Ideas that Have Shaped Our World View,* Richard Tarnas, New York: Ballantine Books, 1991.

[cxiv] *The Birth of Modern Science*, Paolo Rossi, 2001, Blackwell Publishing. ISBN 0631227113.

[cxv] *Leonardo da Vinci: Italian Artist, Engineer and Scientist,* Encyclopedia Britannica. https://www.britannica.com/biography/Leonardo-da-Vinci.

[cxvi] *Leonardo da Vinci, Artist, Inventor and Universal Genius of the Renaissance,* http://www.leonardo-history.com/life.htm?Section=S6; http://www.history.com/topics/leonardo-da-vinci.

[cxvii] *The life and Times of Leonardo,* Liana Bartolon, 1967, Paul Aamlyn, London, ISBN 075251587X.

[cxviii] *The Passion of the Western Mind: Understanding the Ideas that Have Shaped Our World*, Richard Tarnas, New York: Ballantine Books, 1991.

[cxix] *Landmarks in Western Science: From Prehistory to the Atomic Age*, Peter Whitfield, New York: Rutledge, 1999.

[cxx] *The Search for the Beginning and End of the Universe*, John Seife, New York: Penguin Books, 2003.

[cxxi] *The Whole Shebang: A State-of-the-Universe(s) Report*, Timothy Ferris, New York: A Touchstone Book, 1998.

[cxxii] *The Passion of the Western Mind: Understanding the Ideas that Have Shaped Our World*, Richard Tarnas, New York: Ballantine Books, 1991.

[cxxiii] Ibid.

[cxxiv] *Robert Hooke,* Encyclopedia.com, Complete Dictionary of Scientific Biography, Charles Scribner's Sons, 2008.

[cxxv] *Landmarks in Western Science: From Prehistory to the Atomic Age*, Peter Whitfield, New York: Rutledge, 1999.

[cxxvi] *A Brain for all Seasons: Human Evolution & Abrupt Climate Change,* William H. Calvin, Chicago: The University of Chicago Press, 2002.

[cxxvii] *Wisdom of the West,* Bertrand Russell, London: Bloomsbury Books, 1989.

[cxxviii] *What is Enlightenment?,* Immanuel Kant, In "The Portable Enlightenment Reader". Edited by Isaac Krannick, New York, Penguin Books, 1995.

[cxxix] *The Whole Shebang: A State-of-the-Universe(s) Report*, Timothy Ferris, New York: A Touchstone Book, 1998.

[cxxx] *Thinking Whole,* Larry Bell, Stairway Press, 2018.

[cxxxi] *The Enlightenment,* Josh Rahn, The Literature Workshop, http:www.online-literature.com, 2011.

[cxxxii] *The Unity of Knowledge,* Edward O. Wilson, New York: Alfred A. Knopf, 1998.

[cxxxiii] *Wisdom of the West,* Bertrand Russell, London: Bloomsbury Books, 1989.

[cxxxiv] *Out of Chaos: Evolution from the Big Bang to Human Intellect*, Wayne Bundy, Boca Raton: Universal Publishers, 2007.

[cxxxv] *The Unbound Prometheus,* David S. Landes, Cambridge, Massachusetts: Press Syndicate of the University of Cambridge, 1969, ISBN 978-0-521-09418-4.

[cxxxvi] *English and American Tool Builders,* Joseph Wickham Roe, New Haven, Connecticut, Yale University, 1916, Reprinted by

Mcgraw-Hill, New York and London, 1926 and by Lindsay
Publications, Bradley, Illinois.

[cxxxvii] *The Wealth and Poverty of Nations,* David Landes, 1999, New
York: W.W. Norton and Company.

[cxxxviii] *Empire of Cotton: A Global History,* Sven Beckert, 2014, New
York: U.S. Vintage Books Division of Penguin Random House.

[cxxxix] *English and American Tool Builders,* Joseph Wickham Roe,
New Haven, Connecticut, Yale University, 1916, Reprinted by
Mcgraw-Hill, New York and London, 1926 and by Lindsay
Publications, Bradley, Illinois.

[cxl] *Inventing the Cotton Gin: Machine and Myth in Antebellum
America*, Angela Lakwete, 2005, Baltimore: John Hopkins University
Press.

[cxli] *The Unbound Prometheus,* David S. Landes, Cambridge,
Massachusetts: Press Syndicate of the University of Cambridge, 1969,
ISBN 978-0-521-09418-4.

[cxlii] *Technological Transformations and Long Waves*, Robert Ayres,
1989, PDF.

[cxliii] *A History of Industrial Power in the United States*, Louis
Hunter, 1985, Vol. 2: Steam Power, Charlottesville: University Press
of Virginia. .

[cxliv] *A History of Metallurgy, Second Edition*, R.F. Tylecote, 1992:
London: Maney Publishing, for the Institute of Materials.

[cxlv] *The Unbound Prometheus,* David S. Landes, Cambridge,
Massachusetts: Press Syndicate of the University of Cambridge, 1969,
ISBN 978-0-521-09418-4.

[cxlvi] *The Most Powerful Idea in the World: A Story of Steam
Industry and Invention*, William Rosen, 2012, Chicago: University of
Chicago Press.

[cxlvii] *The Unbound Prometheus,* David S. Landes, Cambridge,
Massachusetts: Press Syndicate of the University of Cambridge, 1969,
ISBN 978-0-521-09418-4.

[cxlviii] *From the American System to Mass Production, 1800-1932:
The Development of Manufacturing Technology in the United States*,
David A. Hounshell, Baltimore: Johns Hopkins University Press.

[cxlix] *Lectures on Economic Growth*, Robert Lucas, 2002, Cambridge: Harvard University Press.

[cl] *Technological Transformations and Long Waves*, Robert Ayres, 1989, PDF.

[cli] *Sapiens: A Brief History of Humankind*, Yuval Noah Harari, New York, London, Toronto, Sydney, New Delhi, Auckland: HarperCollins/Harper Perennial, 2015 .

[clii] *Hyperspace: A Scientific Odyssey through Parallel Universes, Time Warps, and the 10th Dimension*, Michio Kaku, 1994, New York: Anchor Books, Doubleday.

[cliii] *An Encyclopedia of the History of Technology,* Ian McNeil, 1990, London: Routledge. ISBN 0-415-14792-1.

[cliv] *The Education of Thomas Edison,* Jim Powell, February 1, 1995, Foundation for Economic Education.

[clv] Ibid.

[clvi] Ibid.

[clvii] Ibid.

[clviii] *Thinking Whole: Rejecting Half-Witted Left & Right Brain Limitations*, Larry Bell, 2018, Arizona: Stairway Press ISBN 978-1-949267-02-0.

[clix] *S-Matrix Interpretation of Quantum Theory,* Henry Stapp, June 22, 1970, Lawrence Berkeley Laboratory.

[clx] *The Evolution of Physics,* Albert Einstein and Leopold Infeld, 1961, Simon and Shuster.

[clxi] *Thinking Whole: Rejecting Half-Witted Left & Right Brain Limitations*, Larry Bell, 2018, Arizona: Stairway Press ISBN 978-1-949267-02-0.

[clxii] *The Dancing Wu Li Masters: An Overview of the New Physics*, Gary Zukav, 2001, HarperCollins.

[clxiii] *Taking Flight: Inventing the Aerial Age, from Antiquity through the First World War*, Richard P. Hallion, 2003, New York: Oxford University Press ISBN 0195160355.

[clxiv] *The Practical Significance of Konstantin Tsilokovsky's Proposals in the Field of Rocketry,* Sergei P. Korolev, 1986 paper presented at a meeting commemorating Tsiolkovsky's 100th birthday, Sept. 17,

1957, USSR Academy of Sciences, Moscow, reprinted in English in History of the USSR; New Research, 5, Social Sciences Today, Moscow.

[clxv] *Creative Legacy of Academician Sergi Pavlovich Korolev,* M.V. Keldysh, editor, and G.S. Vetrov, compiler,(in Russian), Moscow, Nauka, 1980.

[clxvi] Ibid.

[clxvii] Ibid.

[clxviii] Ibid.

[clxix] Ibid.

[clxx] *Recollections of Childhood: Early Experiences in Rocketry as Told by Wernher von Braun,* 1963, MSFC History Office, NASA Marshall Space Flight Center. http://biography.com/people/wernher-von-braun-9224912.

[clxxi] Ibid.

[clxxii] Ibid.

[clxxiii] Ibid.

[clxxiv] Ibid.

[clxxv] Ibid.

[clxxvi] Ibid.

[clxxvii] Ibid.

[clxxviii] Ibid.

[clxxix] *All Quiet on the Western Front,* Erich Maria Remarque, 1929, Ballantine Books.

[clxxx] *Flops of Famous Inventions,* George L. Dowd, Popular Science, 1930.

[clxxxi] *Helicopter Development in the Early Twentieth Century,* Judy Rumerman, Centennial Flight Commission. https://www.centennialofflight.net/essay/Rotary/early_20th_century/HE2.htm.

[clxxxii] *Kaman K-225,* R.D. Connor and R.E. Lee, July 21, 2001, Smithsonian National Air and Space Museum.

[clxxxiii] BBC on this Day, 15, 1940: *Victory for RAF in Battle of Britain,* September 15, 2005, UK: BB.

[clxxxiv] The Pacific War Online Encyclopedia.

[clxxxv] *Eisenhower on the Opportunity Cost of Defense Spending,* November 12, 2007, as reported by Scott Horton in Harper's Magazine, March 2, 2019.

[clxxxvi] *Stalin's Wars From World War to Cold War, 1939-1953,* Geoffrey Roberts, 2006, Cambridge, Mass: Yale University Press, ISBN 9780300112047.

[clxxxvii] *The Moldavians: Romania, Russia, and the Politics of Culture,* Charles King, 2000, Hoover Institution Press, ISBN 9780817997922.

[clxxxviii] Red Century: From the 'October Revolution' in 1917, Communism swept the Globe, Will Englund, October 26, 2016, The Washington Post.

[clxxxix] *Nuclear Forces Reduced While Modernizations Continue, Says SIPRI,* June 16, 2014, Stockholm International Peace Research Institute.

[cxc] *Tass,* September 8, 1958.

[cxci] *Corona: America's First Spy Satellite Program,* Quest, Wayne Day, Grand Rapids, MI: Cspace Press, Summer, 1995.

[cxcii] *US Reconnaissance Satellite Programs,* Jonathan McDowell and Robert A. McDonald, *Corona Success for Space Reconnaissance,* PE & RS Photogrammetric Engineering and Remote Sensing, June 1995.

[cxciii] *Pravda,* November 29, 1977.

[cxciv] *Tass,* September 8, 1958.

[cxcv] National Intelligence Estimate, The Russian Space Program, Dec.5, 1962, as reported in Hartford, James, *Korolev: How One Man Masterminded the Soviet Drive to Beat America to the Moon,* Wiley & Sons, Inc., 1997.

[cxcvi] *The Practical Significance of Konstantin Tsilokovsky's Proposals in the Field of Rocketry,* Sergi P. Korolev, paper presented at a meeting commemorating Tsiolkovsky's 100th birthday, Sept. 17, 1957, USSR Academy of Sciences, Moscow, reprinted in English in History of the USSR; New Research, 5, Social Sciences Today, Moscow, 1986.

[cxcvii] *The Way it Was: the Difficult Fate of the N-1 Project,* M. Rebrov, Krasnaya Zvezda (Russian), Jan. 15, 1990, No.11, p.4, as reported in Hartford, James, *Korolev: How One Man Masterminded*

the Soviet Drive to Beat America to the Moon, Wiley & Sons, Inc., 1997.

[cxcviii] Ibid.

[cxcix] *SP-4209 The Partnership: A History of the Apollo-Soyuz Test Project,* NASA Space History Office, https://history.nasa.gov/SP-4209/ch2-4.htm.

[cc] *There are not enough resources to support the world's population*, John Guillebaud, June 10, 2014, https://www.abc.net.au/radionational/programs/ockhamsrazor/there-are-not-enough-resources-to-support-the-worlds-population/5511900.

[cci] *The Global Exploration Roadmap*, International Space Exploration Coordination Group, NASA Headquarters, January 2018. http://www.globalspacexloration.org.

[ccii] *No Dream is Too High*, Buzz Aldrin with Ken Abraham, National Geographic, 2016.

[cciii] *Time Enough for Love*, Robert A. Heinlein, *GP Putman's Sons*, 1973.

[cciv] *Altered States: Self-experiments in Chemistry*, Oliver Sacks, *The New Yorker*, August 2012.

[ccv] *Social Animal*, David Brooks, *The New Yorker, Annals of Psychology*, January 17, 2011.

[ccvi] *The Hedgehog and the Fox*, Isiah Berlin, *Wiedenfeld & Nicholson*, 1953.

[ccvii] *What Makes a Renaissance Man?*, Olivia Goldmill, *The Telegraph,* December 2014.

[ccviii] *I Sing the Body's Pattern Recognition Machine*, Diane Ackerman, June 15, 2004, New York Times.

[ccix] *Perception in Chess*, Chase and Simon, Cognitive Psychology, 1973.

[ccx] *Holism and Evolution*, Jan Smuts, Greenwood Press, 1973.

[ccxi] *General System Theory: Fundamentals, Development, Applications*, Ludwig von Betlanffy, New York, George Braziller, 1968.

[ccxii] *Introduction to Cybernetics*, Ross Ashby, Routledge Kegan and Paul, 1964.

[ccxiii] *Industrial Dynamics-A Major Breakthrough for Decision Makers, Jay Forrester*, Harvard Business Review, 1958.

[ccxiv] *Time Enough for Love*, Robert A. Heinlein, GP Putman's Sons, 1973.

[ccxv] *Einstein, His Life and Universe: the Basis for Genius*, Walter Isaacson, Simon and Schuster Paperbacks, 2017.

[ccxvi] Ibid.

[ccxvii] *Einstein as a Student* (Unpublished paper), Dudley Herschbach, Provided to the Author, 2005.

[ccxviii] *Einstein, His Life and Universe: The Basis for Genius*, Walter Isaacson, Simon and Schuster Paperbacks, 2017.

[ccxix] Ibid.

[ccxx] Ibid.

[ccxxi] *My Inventions: The Autobiography of Nikola Tesla*, http://www.teslaautobiography.com, 2018.

[ccxxii] *Creativity*, http://psychology.wikia.com/wiki/Creativity, 2018.

[ccxxiii] *The Evolution of Imagination*, Stephen T. Asma, University of Chicago Press, 2017.

[ccxxiv] Ibid.

[ccxxv] *Flow and the Psychology of Discovery and Invention*, Mihaly Csikzentmihalyi, Harper Collins, 1996.

[ccxxvi] *The Pursuit of happiness: Bringing the Science of Happiness to Life*, Mihaly Csikszentmihalyi, 2018.

[ccxxvii] *The Evolution of Imagination*, Stephen T. Asma, *University of Chicago Press*, 2017.

[ccxxviii] *The Age of Insight: The Quest to Understand the Unconsciousness in Art, Mind, and Brain: from Vienna 1900 to the Present*, Eric Kandel, Random House, 2012.

[ccxxix] *The Evolution of Imagination*, Stephen T. Asma, University of Chicago Press, 2017.

[ccxxx] *Descent of Man*, Charles Darwin, Penguin Classics, 2004.

[ccxxxi] *The Evolution of Imagination*, Stephen T. Asma, University of Chicago Press, 2017.

[ccxxxii] *Affective Neuroscience: The Foundations of Human and Animal Emotions*, Jaak Panksepp, Oxford: Oxford University Press, 2004.

[ccxxxiii] *The Evolution of Imagination*, Stephen T. Asma, University of Chicago Press, 2017.

[ccxxxiv] Ibid.

[ccxxxv] Ibid.

[ccxxxvi] *The Runaway Species: How Human Creativity Remakes the World*, Anthony Brandt and David Eagleman, Canongate Books Ltd., Great Britain, 2017.

[ccxxxvii] Ibid.

[ccxxxviii] Ibid.

[ccxxxix] *Thinking Fast and Slow*, Daniel Kahneman, Farrar, Strauss and Giroux, 2011.

[ccxl] *On Understanding Nonliteral Speech: Can People Ignore Metaphors?*, Sam Glucksberg, Patricia Gildea, and Howard G. Bookin, Journal of Verbal Learning and Visual Behavior, 1982.

[ccxli] *Thinking Fast and Slow*, Daniel Kahneman, Farrar, Strauss and Giroux, 2011.

[ccxlii] Ibid.

[ccxliii] Ibid.

[ccxliv] Ibid.

[ccxlv] Ibid.

[ccxlvi] Ibid.

[ccxlvii] Ibid.

[ccxlviii] *Ego Depletion and the Strength Model of Self-Control: A Meta-Analysis*, Martin S. Hagger et al., Psychological Bulletin, 2010.

[ccxlix] *Thinking Fast and Slow*, Daniel Kahneman, Farrar, Strauss and Giroux, 2011.

[ccl] Ibid.

[ccli] *The Runaway Species: How Human Creativity Remakes the World*, Anthony Brandt and David Eagleman, Canongate Books Ltd., Great Britain, 2017.

[cclii] Ibid.

[ccliii] *Engineering of Jihad: The Curious Connection between Violent Extremism and Education*, Diego Gambetta and Steffen Hertog, Princeton University Press, 2016.

[ccliv] *Free Play: Improvisation in Life and Art*, Stephen Nachmanovitch, Jeremy Thacher/Putnam, 1990.

[cclv] *Edison: A Life of Invention*, Paul Israel, John Wiley, 1998.

[cclvi] *The Runaway Species: How Human Creativity Remakes the World*, Anthony Brandt and David Eagleman, Canongate Books Ltd., Great Britain, 2017.

[cclvii] *No Innovator's Dilemma Here: in Praise of Failure*, James Dyson, Wired, April 8, 2011.

[cclviii] *The Runaway Species: How Human Creativity Remakes the World*, Anthony Brandt and David Eagleman, Canongate Books Ltd., Great Britain, 2017.

[cclix] *55 Quotes to Inspire Creativity, Innovation and Action*, Psychology Today, 2010.

[cclx] Linus Pauling Quotes, BrainyQuote.com. Xplore Inc, April 2, 2018. https://www.brainyquote.com/quotes/linus_pauling_163645.

[cclxi] *The Runaway Species: How Human Creativity Remakes the World*, Anthony Brandt and David Eagleman, Canongate Books Ltd., Great Britain, 2017.

[cclxii] *My Inventions: The Autobiography of Nikola Tesla*, http://www.teslaautobiography.com, 2018.

[cclxiii] *Talent is made, not born: Is innate intelligence highly over-rated in our society?* InpaperMagazine, ThriveWork Staff, Dawn News, January 15, 2011.

[cclxiv] *Talent is Overrated: What Really Separates World-Class Performers from Everybody Else,* Geoff Colvin, Penguin, 2008.

[cclxv] *What it takes to be great: Research now shows that the lack of natural talent is irrelevant to great success. The secret? Painful and demanding practice and hard work*, Geoffrey Colvin, FortuneMagazine.com, October 19, 2006.

[cclxvi] *Where machines could replace humans- and where they can't (yet)*, Michael Chui, James Manyika, and Mehdi Miremadi, McKinsey Quarterly, July 2016.

[cclxvii] *The History of Computing is both Evolution and Revolution*, Justin Zobel, The Conversation.com, May 31, 2016.

[cclxviii] Ibid.

[cclxix] *Ken Olsen: Did Digital founder Ken Olsen say there was "no reason for any individual to have a computer in his home?"*, Snopes.com, https://www.snopes.com/fact-check/ken-olsen/.

[cclxx] *Why GANs give artificial intelligence wonderful (and scary) capabilities*, Gabriel Sidhom, August 23, 2017, Orangesv.com.

[cclxxi] *Quantum Computing May Outsmart All Cyber Defenses*, Larry Bell, Newsmax, November 12, 2017.

[cclxxii] *10 Breaktrhough Technologies*, MIT Technology Review, 2018

[cclxxiii] *IBM's Quantum-Computing Service Now Has More Than 100 Clients*, Sara Castellanos, January 9, 2020, Wall Street Journal.

[cclxxiv] *Quantum Computing Leap Still Falls Short*, Asa Fitch and Aaron Tilley, January 7, 2020, Wall Street Journal.

[cclxxv] *Algorithms with Minds of Their Own*, Curt Levy and Ryan Hagemann, Wall Street Journal, November 13, 2017.

[cclxxvi] *Post-quantum cryptology—dealing with the fallout of physics success*, Daniel J. Bernstein and Tanja Lange, National Science Foundation/European Commission, April 9, 2017.

[cclxxvii] *Reinventing Ourselves: How Technology is Rapidly and Radically Transforming Humanity*, Larry Bell, 2019, Stairway Press.

[cclxxviii] *Industrial robots will replace manufacturing jobs—and that's a good thing*, Matthew Randall, TechCrunch.com, October 9, 2016.

[cclxxix] *Jobs lost, jobs gained: What the future of work will mean for jobs, skills and wages*, McKinsey & Company, James Manyika, Susan Lund, Michael Chui, Jacques Bughin, Jonathan Woetzel, Paul Batra, Ryan Ko, and Saurabh Sanghvi, November, 2017.

[cclxxx] Ibid.

[cclxxxi] *Every study we could find on what automation will do to jobs, in one chart"*, Erin Winick, MIT Technology Review.com, January 25, 2018.

[cclxxxii] *The Impact of the Internet on Society: A Global Perspective,* Manuel Castells, September 8, 2014, Technology Review.com (Chair, Professor of Communication Technology at the University of Southern California, Los Angeles).

[cclxxxiii] *What Makes Human's Different? Fiction and Cooperation,* Arik Gabbi, Smithsonian Magazine, February 2015.

[cclxxxiv] *Information Technology and Moral Values,* Stanford Encyclopedia of Philosophy, June 12, 2012.

[cclxxxv] *Artificial Intelligence: Will It Destroy Mankind?* Lee Gruenfeld, Newsmax Magazine, November 2019.

[cclxxxvi] *What Makes Human's Different? Fiction and Cooperation,* Arik Gabbi, Smithsonian Magazine, February 2015.

[cclxxxvii] *What Happens When Artificial Intelligence Turns On Us?,* Erica R. Hendry, January 21, 2014, Smithsonian.com.

[cclxxxviii] Ibid.

[cclxxxix] *Artificial Intelligence: Will It Destroy Mankind?* Lee Gruenfeld, Newsmax Magazine, November 2019.

[ccxc] *Technology isn't just changing society—it's changing what it means to be human: A conversation with historian of science Michael Bess,* Sean Illing, February 23, 2018, Vox.com.

[ccxci] *Phaedrus,* section 275d

[ccxcii] *Technology isn't just changing society—it's changing what it means to be human: A conversation with historian of science Michael Bess",* Sean Illing, February 23, 2018, Vox.com.

[ccxciii] Ibid.

[ccxciv] Ibid.

[ccxcv] *Information Technology and Moral Values,* Stanford Encyclopedia of Philosophy, June 12, 2012.

[ccxcvi] Ibid.

[ccxcvii] *On the Intrinsic Value of Information Objects and the Infosphere,* L. Floridi, 2003, Ethics and Information Technology, 4(4): 287-304.

[ccxcviii] *The Relentless Pace of Automation,* David Rotman, February 13, 2017, Technology Review.

[ccxcix] *The Human Promise of the Revolution,* Kai-Fu Lee, September 15-16, Wall Street Journal.

[ccc] Ibid.

[ccci] *The ethics of Artificial Intelligence,* Justin Lee, June 26, 2018, GrowthBot.org.

[cccii] *Pence Cautions Google on China,* Michael C. Bender and Dustin Volz, October 5, 2018. Wall Street Journal.

[ccciii] *Beijing Expands Its Cybersecurity Regulations,* Shan Li, October 6-7, 2018, Wall Street Journal.

[ccciv] *From Falun Gong to Xinjiang: China's Repression Maestro,* Chun Han Wong, April 4, 2019, Wall Street Journal.

[cccv] *Gifts That Snoop? The Internet of Things Is Wrapped in Privacy Concerns,* Bree Fowler, December 13, 2017, Consumer Reports.

[cccvi] *From Falun Gong to Xinjiang: China's Repression Maestro,* Chun Han Wong, April 4, 2019, Wall Street Journal.

[cccvii] *Code: And Other Laws of Cyberspace, Version 2.0/Edition 2,* Lawrence Lessig, December 28, 2006, Basic Books.

[cccviii] *They Are Watching You,* Robert Draper, February 2018, National Geographic.

[cccix] Ibid.

[cccx] *The Imaginary Real World of Cybercities,* M.C. Boyer, 1992, Assemblage.

[cccxi] *A model Korean ubiquitous eco-city? The politics of making Songdo,* S.T. Shwayri, 2013, Journal of Urban Technology.

[cccxii] *They Are Watching You,* Robert Draper, February 2018, National Geographic.

[cccxiii] Ibid.

[cccxiv] Ibid.

[cccxv] Ibid.

[cccxvi] *Welcome to the Quiet Skies,* Jana Winter, July 28, 2018, the Boston Globe.

[cccxvii] Ibid.

[cccxviii] *We let technology into our lives. And now it's starting to control us,* Rachel Holmes, November 28, 2016, The Guardian.com.

[cccxix] *ARTIFICIAL INTELLIGENCE: The Worlds that AI Might Create,* Michael Totty, October 14, 2019, Wall Street Journal Report.

[cccxx] *Stephen Hawking: Artificial Intelligence Could Wipe Out Humanity When it Gets Too Clever as Humans Will be Like Ants,* Andrew Griffin, October 8, 2015, Independent.com.

[cccxxi] *Lights Out: A Cyberattack, A Nation Unprepared, Surviving the Aftermath,* Ted Koppel, 2015, Broadway Books, an imprint of the Crown Publishing Group, a division of penguin Random House, LLC.

[cccxxii] *Cyber War,* Richard Clarke and Robert K. Knake, 2010, HarperCollins.

[cccxxiii] *The Perfect Weapon,* David E. Sanger, 2018, Crown Publishing Group, Penguin Random House LLC, New York.

[cccxxiv] Ibid.

[cccxxv] *The Fifth Domain: Defending Our Country, Our Companies, and Ourselves in the Age of Cyber Threats,* Richard A Clarke and Robert K. Knake, 2019, New York, Penguin Press.

[cccxxvi] *The Perfect Weapon,* David E. Sanger, 2018, Crown Publishing Group, Penguin Random House LLC, New York.

[cccxxvii] *Cyber War,* Richard Clarke and Robert K. Knake, 2010, HarperCollins.

[cccxxviii] Ibid.

[cccxxix] *FERC Requires Expanded Cybersecurity Incident Reporting,* Federal Energy Regulatory Commission, July 19, 2018, www.ferc.gov/media/news-releases/2018/2018-3/07-19-18-E-1.asp.

[cccxxx] *The Fifth Domain: Defending Our Country, Companies, and Ourselves in the Age of Cyber Threats,* Richard A. Clarke and Robert K Knake, 2019, Penguin Press.

[cccxxxi] *The 'cyberwar' era began long ago,* Ron Kelson, Pierluigi Paganini, Benjamin Gittins, and David Pace, June 25, 2012, The Malta Independent Online.

[cccxxxii] *The Fifth Domain: Defending Our Country, Our Companies, and Ourselves in the Age of Cyber Threats,* Richard A Clarke and Robert K. Knake, 2019, New York, Penguin Press.

[cccxxxiii] *Cyber War,* Richard Clarke and Robert K. Knake, 2010, HarperCollins.

[cccxxxiv] *The 'cyber war' era began long ago,* Ron Kelson, Pierluigi Paganini, Benjamin Gittins, and David Pace, June 25, 2012, The Malta Independent Online.

[cccxxxv] *The Perfect Weapon,* David E. Sanger, 2018, Crown Publishing Group, Penguin Random House LLC, New York.

[cccxxxvi] *Nuclear Weapons and Foreign Policy*, Henry Kissinger, 1957, Council on Foreign Relations, Published by Harper & Row, Inc. ISBN 978-0-393-00494-6.

[cccxxxvii] *Finally, a Presidential EMP Order that may Save American Lives,* Peter Pry, March 28, 2019, The Hill.

[cccxxxviii] *New EMP Warning: US will 'Cease to Exist,' 90 Percent of Population will Die,* Paul Bedard, January 24, 2019, Washington Examiner.

[cccxxxix] Ibid.

[cccxl] *Trump Moves to Protect America from Electromagnetic Pulse Attack,* Ariel Cohen, April 5, 2019, Forbes.

[cccxli] *Beyond Footprints and Flagpoles,* Buzz Aldrin and Larry Bell (Manuscript in progress).

[cccxlii] Ibid.

[cccxliii] Ibid.

[cccxliv] Ibid.

[cccxlv] Ibid.

[cccxlvi] Ibid.

[cccxlvii] Ibid.

[cccxlviii] Ibid.

[cccxlix] *No Dream is Too High,* Buzz Aldrin with Ken Abraham, *National Geographic,* 2016.

[cccl] *Beyond Footprints and Flagpoles,* Buzz Aldrin and Larry Bell (Manuscript in progress).

[cccli] *The Runaway Species: How Human Creativity Remakes the World,* Anthony Brandt and David Eagleman, Canongate Books Ltd., Great Britain, 2017.

www.ingramcontent.com/pod-product-compliance
Lightning Source LLC
Chambersburg PA
CBHW031230090426
42742CB00007B/145